Environments of Empire

FLOWS, MIGRATIONS, AND EXCHANGES

Mart A. Stewart and Harriet Ritvo, editors

The Flows, Migrations, and Exchanges series publishes new works of environmental history that explore the cross-border movements of organisms and materials that have shaped the modern world, as well as the varied human attempts to understand, regulate, and manage these movements.

Environments of Empire

Networks and Agents of Ecological Change

EDITED BY

Ulrike Kirchberger and Brett M. Bennett

The University of North Carolina Press CHAPEL HILL

Set in Merope Basic by Westchester Publishing Services
Manufactured in the United States of America

The University of North Carolina Press has been a member
of the Green Press Initiative since 2003.

Library of Congress Cataloging-in-Publication Data
Names: Kirchberger, Ulrike, editor. | Bennett, Brett M., 1983– editor.
Title: Environments of empire : networks and agents of ecological change / edited by Ulrike Kirchberger and Brett M. Bennett.
Other titles: Flows, migrations, and exchanges.
Description: Chapel Hill : The University of North Carolina Press, [2020] | Series: Flows, migrations, and exchanges | Includes bibliographical references and index.
Identifiers: LCCN 2019035098 | ISBN 9781469655925 (cloth : alk. paper) | ISBN 9781469655932 (paperback : alk. paper) | ISBN 9781469655949 (ebook)
Subjects: LCSH: Human ecology—History—19th century—Case studies. | Human ecology—History—20th century—Case studies. | Global environmental change—History—19th century—Case studies. | Global environmental change—History—20th century—Case studies. | Imperialism—History. | Environmental sciences—History. | Europe—Colonies—History. | Turkey—History—Ottoman Empire, 1288–1918.
Classification: LCC GF50 .E57 2020 | DDC 304.2094/09034—dc23
LC record available at https://lccn.loc.gov/2019035098

Cover illustration: From Alphonse Milne Edwards and Henri Milne Edwards, *Recherches pour servir à l'histoire naturelle des mammifères*, tome second-atlas (Paris: G. Masson, 1868–1874). Courtesy of the Biodiversity Heritage Library.

Contents

Tables

Acknowledgments

The chapters in this volume comprise a selection of papers that were first presented at an international conference at the University of Kassel, Germany, in September 2015. We are grateful to the University of Kassel for hosting and funding the conference; our thanks likewise to all participants for their contributions and for making the meeting an interesting and enjoyable experience. We would also like to thank Harriet Ritvo and Mart Stewart for accepting the manuscript for publication in the series Flows, Migrations, and Exchanges. At UNC Press, Brandon Proia guided us through the publication process. Michelle Witkowski and a team of highly competent editors prepared the manuscript for publication. The reports of two anonymous readers helped us to improve the quality of the volume. At the University of Kassel, Austeja Schumann, Miguel Ohnesorge, Nhat Luong, and Lion Arendt provided valuable assistance.

Environments of Empire

Introduction

ULRIKE KIRCHBERGER

The ecological dynamics and impacts of European imperialism have received significant attention from historians since the publication of Alfred Crosby's influential book *Ecological Imperialism* in 1986.[1] It is now well known that European emigrants initiated long-distance transfers of biota that substantially modified the environments of both colonies and homelands of colonial powers. Global and imperial environmental histories, studies on individual colonies, and global histories of individual species explore the ways imperialism and colonization transformed the natural environments in different parts of the world.[2]

There are, however, still significant knowledge gaps that limit our ability to understand the patterns, causes, and consequences of this process. The field lacks comparative and transnational analyses examining the environmental histories of two or more European empires. The majority of research focuses on one imperial setting, the most frequently studied being the British Empire.[3] New literature that takes a comprehensive European approach, such as the monograph by Corey Ross, follows the core–periphery perspective of European imperialism and highlights how Europeans exploited the natural resources of the tropical world.[4] New collected volumes sometimes pursue a multidirectional approach and show how transnational and multispecies networks crisscrossed colonial landscapes; but again, they tend to focus on the British Empire.[5] Others deal with ecological transfers and politics between twentieth-century nation states.[6] Most of these studies have neglected Islamic contexts and regions.

This volume advances research on the causes and dynamics of ecological change in imperial contexts in several respects. It offers a wider comparative and transnational perspective. Nine case studies focus on the Ottoman, French, British, Dutch, and German empires and analyze sources in Turkish, Arabic, German, Dutch, French, and English. They explore hitherto neglected empires and colonies, such as the Ottoman Empire and German West Africa, alongside imperial locations that have received more attention by Anglo-American environmental historians, such as British India, Australia, and New Zealand. Some chapters are comparative and transnational in themselves and deal, for example, with the Ottoman and Western European empires, with the transfer of agronomic techniques between the "old empires" in the Atlantic world and the "new territories" in Africa, and with a zoological network that connected Switzerland, the Netherlands, and West Africa. Other chapters focus on one region or empire, but they also establish transimperial connections and invoke comparisons for their case studies.

Furthermore, the volume addresses important methodological problems that concern the question of nonhuman agency and the ways in which historians can use insights from nonhistorical disciplines. Some chapters experiment with approaches from Actor-network theory (ANT) and the emergent field of animal studies, which question anthropocentric perspectives and interpret animals as actors that are endowed with their own specific agency. Other chapters introduce French and German social history concepts, such as the public sphere and public scandalization, into environmental history to shed new light on the dynamics that shaped ecological transfers in colonial contexts. The epilogue explores how scientific insights can be used to corroborate or challenge historical interpretations of ecological change.

The book is divided into three sections. Each section analyzes a key dimension of imperial ecological change. The first three chapters concentrate on how nation-building shaped ecological transfer. The rise of the national public sphere, the formation of the nation state, and the desire of France and Germany to economically exploit the non-European world provide the backgrounds for explaining transcontinental species transfer. In the second section, the chapters examine scientific institutions and personalities. They draw on approaches from the history of science, especially work on the circulation and diffusion of knowledge. The chapters in the third section of the volume apply animal studies perspectives and reflect upon the role of animals in transimperial networks.

In the epilogue, Brett Bennett thinks about methodological possibilities beyond network analysis and animal studies. Since these disciplines have often been criticized for neglecting research from the natural sciences when placing the nonhuman in the center of analysis, he discusses how to overcome this deficit and examines how historians can make use of the findings of natural sciences to explain global ecological dynamics and species transfers. By applying new research on invasive species from the ecological sciences, he reassesses Crosby's thesis of the transfer of European species to the non-European world. Whereas the other chapters in the volume concentrate on specific historical case studies, he opens the horizon for broader questions about how to connect historical and scientific thinking about global scales and large periods of time.

Networks

This volume uses the conceptual frame of the network to explore ecological change within and across empires. Network analysis deals with specific relations between different actors and has various meanings across academic disciplines. Ecologists refer to networks to understand multispecies interaction in complex ecosystems.[7] Historians trace patterns of networks in order to go beyond the geographic boundaries of national history, to challenge core–periphery frameworks of imperial history, and to analyze transimperial connections. Sociological network theories, such as ANT, offer different interpretations of knowledge production and transfer, and they examine relations between humans and nonhumans.[8]

The chapters in this volume apply network analysis in different ways. Some of them connect closely with ANT and new research from animal studies in so far as they question anthropocentric hierarchies and investigate hybridities and border-crossings between humans and animals.[9] They deal with horses, hippos, and chimpanzees, but they also show how small and apparently insignificant animals, comparable to the scallops studied by Callon and Latour, sometimes dominated relations and threatened human superiority. Different chapters in this book show how worms, insects, and other small creatures endangered colonial economies.

Other chapters use networks to trace long-distance communications and transfers. They explore how global connections were established that reached far beyond nation states and empires. Each chapter assesses the nature and character of the transimperial and multidirectional networks from a different perspective. Taking on local and global points of view, they show how and by whom transfers were initiated, where exchanges were dense and transfers frequent and where geographical, social, political, and ethnic barriers prevented the flow of knowledge and biota.

The network approach demonstrates that many participants in the transfers were very well connected on the global level. The scientists, in particular, who initiated most of the transfers in this volume, imagined themselves as collaborating with a larger transimperial project of discovery and improvement. Although individuals worked for empires and had national allegiances, they shared common interests and professional norms. They communicated with each other across oceans and imperial borders, exchanged letters, publications, plants, animals, and objects of scientific interest and, in this way, created ecological change.

However, the networks this volume explores did not only consist of informal contacts between individuals; they were also shaped by formal institutions. These provided greater stability and continuity compared to the informal exchanges of scientists, which waxed and waned. Institutions such as colonial forest services, botanical and zoological gardens, natural history museums, agricultural colleges, and colonial research stations were important nodal points that initiated, coordinated, and financed ecological transfers over longer periods of time.[10] The botanical gardens in Buitenzorg and Kew, for example, are relevant in all the chapters of this volume as centers of global botany. Acclimatization societies mobilized animals and plants globally. The acclimatization movement in mid-nineteenth-century France, addressed in Alexander van Wickeren's chapter, pioneered efforts all over the world. Scientists and amateur enthusiasts in France, the Near East, Britain, South Asia, Prussia, and Australia created local acclimatization societies, corresponded with each other and exchanged species.[11] Seedbanks in Damascus and Buitenzorg and journals like *Der Tropenpflanzer,* the *Indian Forester,* and the catalogue of the Natural History Museum in Istanbul, which was sent to other natural history museums in Europe on a regular basis, also contributed to formalizing communication and exchange.

Two chapters in this volume focus on institutions that were nodes in the global networks of ecological change. Semih Celik tells the story of the Natural History Museum that was built in Istanbul in the 1830s and burned down in 1848. He examines the transimperial transfers of knowledge and biota, which developed in the making of the museum, and highlights that these transfers were characterized by mutual exchange and circulation of knowledge between Ottoman and other European contexts of science. He contradicts historians who claim that the transfers were mainly directed from West to East and that they were a sign of Ottoman dependency on Western science.

Florian Wagner deals with Buitenzorg, the botanical garden and research station in Dutch Java. He argues that Buitenzorg was more important in the global circulation of species and agricultural knowledge than historians have acknowledged so far. Buitenzorg was not only one of the most important centers in the global scientific networks in the age of empire, it also played a crucial role in the global shift from the American plantation system to new forms of cash crop production that were practiced in the colonies and protectorates in Africa and Asia beginning in the second half of the nineteenth century. Wagner shows that Buitenzorg pioneered a supposedly "modern" form of agriculture, which was based on scientific research and had overcome the American plantation system that was associated with slavery and exploitation.

Ecological networks, in their informal and formal dimensions, crossed imperial borders in many different ways, and it is an important aim of this volume to examine these transimperial connections. At the same time, however, we do not want to neglect the impact that national policies and processes of state-formation had on ecological networks and transfers. Many chapters in this book show that the ecological policies of imperial governments were characterized by both international cooperation and by the protection of national interests. The administrations in different colonial empires often coordinated their efforts not only to transfer species but also to protect their agricultures from the consequences of unwanted nonhuman migration. Idir Ouahes and Samuel Eleazar Wendt, for example, refer to the locust swarms that regularly invaded north and east Africa. They caused the same damage in French, British, German, and Ottoman spheres of influence. An international bureau to fight locusts was therefore established in Damascus, as Idir Ouahes points out in his chapter.

The role of national identity in the context of species transfer is investigated in Alexander van Wickeren's chapter. He examines how the acclimatization of Cuban tobacco was represented in the rising national public sphere in nineteenth-century France. In the middle of the century, scientists of the French state tobacco monopoly tried to make France independent from Cuban imports by acclimatizing Cuban tobacco varieties in Alsace and the southern Rhine valley. When the acclimatization experiments did not bring the desired results, the presumed failure of the scientists was turned into a national scandal by the emerging mass media.

Samuel Eleazar Wendt, in his chapter on phytopathology in the German colonies in Cameroon and Togo, further explains the ambivalences between national interests and transimperial networking. The German botanists he deals with were part of the global networks of ecological transfer and exchange. Many of them were well connected with the botanical gardens in Buitenzorg. They belonged to a scientific community that exchanged information about germs and pathogens that damaged agricultural production in colonial plantations all over the world. At the same time, however, the German botanists' research served a national agenda. By supporting the plantation economy in Cameroon and Togo, German politicians aimed to overcome dependency on imported raw materials such as rubber, cotton, and cacao. The pathogens that threatened this project were turned into enemies of the German people, and the fight against them was fueled by nationalist propaganda.

Idir Ouahes then turns to the state bureaucracy that was developed to shape the natural environment in French mandate Syria and Lebanon after the First World War. He argues that, in contrast to the preceding Ottoman rule, the era of the French mandate was characterized by direct state intervention to increase the agricultural productivity of Syria and Lebanon. Comparable to the German initiatives in Cameroon and Togo before the First World War, French politicians envisioned a national economy that would benefit from the *mise en valeur* produced in the mandate zones in the Near East. Both the French and German agricultural policies were difficult to realize. They were accompanied by constant struggles against plagues and epidemics caused by insects, germs, and parasites, which ruined the harvests of cotton and other cash crops.

Thus, the volume demonstrates how transimperial networking was related to state formation and government politics. Its chapters show differences, parallels, and crossovers in imperial politics, conservationist, economic, and scientific ambitions. They also deal, in comparative perspective, with the ways perceptions and stereotypes of introduced or invasive species formed in different contexts and changed over time.

Agents

The agents of ecological change in this volume can be categorized into three groups. The first group consists of European colonists, scientists, and travelers; the second one comprises non-Europeans; and the third group encompasses nonhuman agents of transfer, such as animals and plants. In each chapter of this book, the relational dynamics between these three groups are investigated from different perspectives and in different imperial settings.

At first glance, the European scientists who initiated the transfers appear to be most influential. They moved between different imperial locales, participated in multinational expeditions, or served different colonial administrations. Austrian botanists, for example, were plant collectors in the Ottoman Empire and contributed to the making of the Natural History Museum in Istanbul. Scientists from Switzerland worked as experts in the British, French, and Dutch colonies. They acted as important agents of transimperial ecological transfers that have often been overlooked. Their scientific ambitions were closely intertwined with the wider economic interests of imperial expansion. Often, scientists were motivated by the European civilizing mission to "improve" the natural environment in what was perceived to be the imperial periphery.

Whereas the motives of European agents in the ecological networks can be reconstructed without too much difficulty from the written sources they have left, it is far more complicated to trace the interests of the non-European men and women who participated in these networks in considerable numbers. They are, however, present in every chapter in different roles and functions. In Samuel Eleazar Wendt's and Florian Wagner's chapters, Africans and Asians played crucial roles as migrant laborers in the colonial plantation economy. Jodi Frawley, in her chapter on oyster cultivation in the estuaries of Eastern Australia, refers to Chinese workers who lived there in the second half of the nineteenth century and probably introduced Chinese fishing techniques and methods for harvesting oysters. Furthermore, non-European personnel were employed in the botanical gardens and colonial research stations where they did the manual work. In Buitenzorg, as Florian Wagner points out, Javanese students were trained in schools and trial farms to grow cash crops. Equipped with such expertise, they were supposed to spread European agricultural techniques in Southeast Asia.

Non-Europeans, however, were not always the willing assistants of the European civilizing mission. Resistance was common and occurred in different forms. Open rebellion was only the most radical way to express discontent with the European plantation economy. Often, interaction between Europeans and non-Europeans was more complex and forms of protest and ways to protect indigenous ecological traditions were very subtle. Samuel Eleazar Wendt, for example, examines how the German plantation companies in Togo tried to convince the Ewe to give up their traditional ways of growing cotton and to adjust to the needs of the German export-oriented agrobusiness by cultivating cotton species that were introduced by German planters from the Americas. Up to a certain extent, however, the Ewe managed to preserve their own agricultural practices and continued to intersperse their yam and corn fields with cotton plants. Such ambivalences between collaboration, adaptation, and resistance characterized the relations between colonial and indigenous participants in ecological transfers in many respects.

Other forms of ecological transfer between Europeans and non-Europeans took place in the context of research expeditions when African and Asian men and women participated as hunters and collectors and provided local knowledge of the natural environment. Most contributors in this volume emphasize that expeditions, which were initiated by Europeans, could not have been carried out without local participation, neither in the Western European colonial empires nor in the Ottoman Empire.[12] Carey McCormack and Stephanie Zehnle, in particular, focus on European scientific expeditions in South Asia and West Africa. They examine the relations that white male scientists developed with white male non-academics, with non-Europeans, and with nonhumans during their journeys. Carey McCormack deals with Joseph Hooker, the famous director of the Royal Botanic Gardens in Kew, and his journeys in South Asia in the middle of the nineteenth century. Stephanie Zehnle concentrates on the zoological expeditions that Johann Büttikofer, a Swiss biologist, and Franz Xaver Stampfli, a Swiss hunter, undertook in Liberia in the 1880s.

Both chapters show that relations between different groups of agents were complex and changeable and cannot be reduced to a simple dichotomy of powerful European versus subaltern non-European. Carey McCormack highlights that Hooker depended on indigenous botanical knowledge and employed large groups of local people to support him on his journeys. Stephanie Zehnle likewise emphasizes that local knowledge was crucial for the European scientists. She rescues some of the "indigenous experts" from anonymity and introduces the Demerys, a Liberian family whose members were key players in the Swiss-Dutch-West African networks she examines in her chapter. Jackson Demery, who lived with his wife and children on an American Baptist missionary station near Bendu, participated in the expeditions of Büttikofer and Stampfli as a hunter. Demery's daughter Mary collected insects and mollusks for Büttikofer. His son, Archery Thomas Demery, was sent to Leiden to be trained as an animal preparator. After his return to West Africa, he joined Büttikofer and Stampfli as hunter and preparator before dying an early death.

Both chapters not only emphasize that race and gender relations changed in the context of European expeditions, they also examine how the hierarchies of imperial science were contested in the contact zones in Asia and Africa. Network actors that belonged to different social classes, different ethnic groups, and different nonhuman species constantly renegotiated their relations when they met and interacted with each other. By directing the focus on the non-European contact zone, the two case studies not only contribute to current research that complicates gender and racial hierarchies in the European sciences,[13] they also question the master narrative of professionalization and the making of academic disciplines in European academia in the nineteenth century. Although research expeditions in Asia and Africa certainly helped scientists such as Hooker and Büttikofer in their academic careers, both chapters demonstrate that there was no clear-cut teleology of the rise of the white male scientist who established himself at the top of gender, racial, and academic hierarchies in the course of the nineteenth century. The chapters describe the constant rivalries and jealousies between different interest groups, which made it difficult for white male academics to completely exclude amateurs and non-European collaborators from the fame and fortune that could be gained from traveling in Africa and Asia.

Furthermore, the chapters reassess the significance of nonhuman agency in the ecological transfers. Stephanie Zehnle demonstrates that the animals themselves played an important part in the production of knowledge in the West African contact zone and, consequently, in the making of zoology as an academic discipline. She shows how relations between humans and animals changed and how the borders between them sometimes dissolved. Chimpanzees, for example, were regarded as half-humans endowed with human emotions. Putting this idea to the test, Büttikofer convinced (and probably paid) an African woman to raise a newborn chimpanzee together with her own baby.

Jodi Frawley, in her chapter, deals with the mudworms that cohabited in the oysters Australian fishermen imported from New Zealand to cover the rising demand of the booming Australian oyster industries at the end of the nineteenth century. The mudworms in New Zealand oysters spread into Eastern Australian estuaries after having been introduced unintentionally by the fishers. They made the oysters inedible. The oyster industry declined as a consequence. The estuaries are described as contact zones where Europeans and Aboriginal people lived and tried to pursue their economic interests. Jodi Frawley distances negative stereotypes and defines the mudworm as an agent of ecological change. She demonstrates that the unintended introduction of the mudworm not only transformed oyster cultivation practices in the region but changed the ecosystems of the estuaries as a whole.

Nicole Chalmer then turns to the horses that lived in the Esperance Mallee Recherche bioregion on the south coast of Western Australia. Out of all contributors, she applies the animal studies approach most rigorously. This reflects not least in the terminology of her chapter. She refers to the horses in the region as "colonizers" and "landscape engineers." She argues that horses were the cherished partners of human colonizers. European farmers gave them names and behaved very affectionately toward them. However, as wild horses, so-called brumbies, the horses developed their own ways to adjust to and shape the ecosystem of the region. Nicole Chalmer combines research from the historical humanities with findings from the natural sciences and shows how brumbies developed their own infrastructures when they, for example, established pathways to access food and water resources or provided firebreaks through their grazing habits.

The other chapters in the book also show that animals unfolded dynamics that humans did not expect, although they are not based on the animal studies approach to the same degree as the three chapters in the last section. Alexander van Wickeren, Samuel Eleazar Wendt, Idir Ouahes, and Florian Wagner refer to insects, germs, and bacteria that caused unanticipated problems to colonial agriculture. In the same way as mudworms harmed the eastern Australian oyster economy, pathogens damaged plantations in German Cameroon and Togo, germs and rusts affected the acclimatization of Cuban tobacco in France and its colonies, and bacteria, larvae, moths, caterpillars, grasshoppers, and locust populations ruined the harvests in European colonies and protectorates in the Near East and in North and East Africa.

In various ways, from different disciplinary perspectives and in different terminology, all chapters show how easily humans lost control over the transfers they initiated. In contrast to new research, which still pursues anthropocentric perspectives and highlights the powerful impact of Europeans in transforming and exploiting non-European nature and natural resources, this book covers new ground in so far as it consistently questions the self-fashioning of Europeans as the superior re-creators of colonial environments. The volume demonstrates that European scientists, in particular, often realized the limits of ecological engineering and then turned into helpless bystanders who had to observe the unintended consequences of their initiatives. Sometimes they even had to bear public humiliation when their experiments failed.

Chronologies of Ecological Change

Current research on ecological networks and transfers in colonial contexts focuses on spatial aspects. Although global historians grow increasingly aware of questions of periodization and other forms of temporality,[14] the temporal dimension of transimperial species transfer has been much neglected so far. Ecological change, however, has an important temporal quality. To examine chronological structures and situate key trends and developments on the time scale is, therefore, vital for understanding the dynamics of ecological networks and species transfers this volume deals with.

Following the conventions of historical periodization, most chapters in this volume concentrate on transfers and networks in the so-called age of globalization, covering the second half of the nineteenth and early twentieth centuries. This era is commonly associated with processes of acceleration and condensation. Long-distance communication and travel were facilitated by new technologies. In the context of imperial expansion, European time regimes were transferred to other parts of the world.[15] Consequently, transcontinental species transfers could be coordinated more easily and became more frequent. European scientists believed that transferring plants and animals in colonial contexts was part of bringing "progress" and "modernity" to the non-European world. African and Asian forms of ecological knowledge and agricultural practices were shifted back on the timeline to a distant past and a primitive stage of human development.

The "age of globalization" provides a relatively firm, Euro- and anthropocentric time frame ranging from the mid-nineteenth century until the beginning of the First World War. In 1914, international communication networks broke down, transimperial cooperation was interrupted, and, as a result of the war, colonial regimes were substituted by other forms of government. This volume, however, offers further possibilities for the periodization of global ecological change and for defining space/time relations in the context of global species transfer. Florian Wagner, for example, highlights the shift from "old empires" to "new territories," which saw the transition from the "old" plantation economies to a supposedly new, science-based colonial cash crop production. In other chapters, generational patterns play an important role in explaining continuities and changes in the ecological networks of European scientists. Garden directors, zoologists, foresters, and travelers referred to the achievements of their predecessors, quoted their research in their publications, and, by doing so, created continuities of knowledge transfer on the timeline. In this book, such generational transfers can be traced from the Enlightenment travelers in the Ottoman Empire of the late eighteenth century, to scientists in the British Empire in the mid-nineteenth century, to the agronomists in Buitenzorg, to the phytopathologists in the German colonies in West Africa at the end of the nineteenth and early twentieth centuries, and on to the experts in French mandate Syria and Lebanon in the 1920s and 1930s. However, the assumption of a steady and undisturbed teleology of professionalization and specialization over generations of male European scientists is questioned in those chapters that have a narrow time frame and focus on contact zones in the non-European world. They show that hierarchies varied considerably in different spatial units of the global ecological networks and that European scientists were not always in control of the time schedule.

Transcontinental species transfers were shaped by chronological structures that diverged from the time regimes of European imperialism. Non-European calendric systems, and planting and harvesting cycles influenced the organization of the transfers and the ways in which introduced species adjusted to new environments. If we follow ANT and integrate nonhumans as actors, the problem of structuring ecological networks chronologically becomes even more complicated. The biological rhythms of introduced plants and animals often differed from manmade chronology. Clear-cut breaks—such as, for example, the arrival of the First Fleet in Australia or the beginning of the First World War—appear permeable, and they overlap with other time frames that characterized relations between humans and nonhumans in the ecological networks.

Often, the patterns of growth of introduced plants and animals did not rise to the expectations held by scientists and colonial agriculturalists. When introduced populations spread too fast because they had no natural enemies in their new environments, they were soon perceived as a threat to the ecological balance. Just as often, however, plants and animals did not survive the transoceanic journeys or died soon after their arrival. Animals sometimes became sick in their new environment, and plant samples and seeds did not grow at all or too slowly to be used for farming. The speed with which introduced species developed was one of the most important topics in the scientists' correspondences and an important criterion for assessing their economic potential. The discrepancies between the biochronologies of plants and animals and the time regimes of scientists, acclimatizers, and colonial officials involved in the transfers play an important role in many chapters of this volume.

Ecological networks and transfers were thus shaped by different forms of temporality. Which time frame to choose when analyzing ecological change is a matter of perspective and disciplinary approach. In this volume, colonial historians and historians of science prefer clear-cut, rather narrow time frames to analyze ecological change, whereas contributors who concentrate on the role of the animal in historical analysis cover large periods of time. Nicole Chalmer, for example, reaches far back in time to explain the history of the brumbies in Western Australia. Jodi Frawley refers to the "slow violence of the anthropocene" when she explains the transformation in the ecosystems of the eastern Australian estuaries by the oyster industries. When shifting the focus to nonhumans, long-term continuities become more important. They reach far beyond the detailed, small-scale periodization schemes historians need to examine their empirical case studies.

Interdisciplinary tensions with respect to periodization are addressed in Jodi Frawley's chapter and, finally, in Brett Bennett's epilogue, which brings together ecological science and history. Bennett argues for a long-term continuity of a "process of global ecological homogenization" and defines two stages of "ecological globalization"—one reaching from 1500 until 1850, and the other one from the second half of the nineteenth century until today. He deals with the problem of how long-term developments and short-term breaks can be related to each other and invites the reader to think about how directions and asymmetries in global species transfers changed over time.

The volume shows the potential the multidisciplinary approach has for reconceiving ecological change in the context of imperial expansion. In new comparative perspectives, it assesses human and nonhuman exchange both on the global level across empires and world regions and on the local level where different groups of participants interacted with each other. From different methodological points of view, the chapters show how relations between humans and nonhumans changed and how hierarchies shifted in the process of ecological transfers. In this way, the volume aims to shed new light on the classical topic of "ecological imperialism."

Notes

1. Alfred W. Crosby, *Ecological Imperialism: The Biological Expansion of Europe, 900-1900* (New York: Cambridge University Press, 1986).

2. See, to name but a few examples out of a growing body of literature, Lucile Hunt Brockway, *Science and Colonial Expansion: The Role of the British Royal Botanic Gardens* (New Haven, CT: Yale University Press, 2002, first published by Academic Press in 1979); Richard Grove, *Green Imperialism: Colonial Expansion, Tropical Island Edens and the Origins of Environmentalism, 1600–1860* (Cambridge: Cambridge University Press, 1995); J. R. McNeill, "Biological Exchange in Global Environmental History," in *A Companion to Global Environmental History*, edited by J. R. McNeill and Erin Stewart Mauldin (London: John Wiley & Sons, 2012), 433–51; James Beattie, *Empire and Environmental Anxiety: Health, Science, Art and Conservation in South Asia and Australasia, 1800–1920* (Basingstoke, UK: Palgrave Macmillan, 2011); John Ryan Fisher, *Cattle Colonialism: An Environmental History of the Conquest of California and Hawai'i* (Chapel Hill: University of North Carolina Press, 2015); Ian Tyrrell, *True Gardens of the Gods: Californian-Australian Environmental Reform, 1860–1930* (Berkeley: University of California Press, 1999); Haripriya Rangan, Judith Carney, and Tim Denham, "Environmental History of Botanical Exchanges in the Indian Ocean World," *Environment and History* 18, no. 3 (2012): 311–42; Ruth Morgan and James Beattie, "Engineering Edens on this 'Rivered Earth'? A Review Article on Water Management and Hydro-Resilience in the British Empire, 1860s–1940s," *Environment and History* 23 (2017): 39–63. For a range of different case studies dealing with transcontinental transfers of species and ecological knowledge, see "Transoceanic Exchanges," a special issue of

Environment and History 21 (2015). For global histories of individual species, see, for example, Tom L. McKnight, *The Camel in Australia* (Melbourne, Australia: Melbourne University Press, 1969); Robin W. Doughty, *The Eucalyptus: A Natural and Commercial History of the Gum Tree* (Baltimore, MD: Johns Hopkins University Press, 2000); William Beinart and Luvuyo Wotshela, *Prickly Pear: The Social History of a Plant in the Eastern Cape* (Johannesburg: Wits University Press, 2011); Brett M. Bennett, "A Global History of Australian Trees," *Journal of the History of Biology* 44 (2011): 125–45.

3. Standard texts on ecological aspects of British imperialism are, for example, John M. McKenzie, *The Empire of Nature: Hunting, Conservation and British Imperialism* (Manchester, UK: Manchester University Press, 1988); Tom Griffiths and Libby Robin, eds., *Ecology and Empire: Environmental History of Settler Societies* (Seattle: University of Washington Press, 1997); William Beinart and Lotte Hughes, *Environment and Empire* (Oxford: Oxford University Press, 2007).

4. Corey Ross, *Ecology and Power in the Age of Empire: Europe and the Transformation of the Tropical World* (Oxford: Oxford University Press, 2017).

5. Jodi Frawley and Iain McCalman, eds., *Rethinking Invasion Ecologies from the Environmental Humanities* (London: Routledge, 2014); James Beattie, Edward Melillo, and Emily O'Gorman, eds., *Eco-Cultural Networks and the British Empire* (London: Bloomsbury, 2014).

6. Erika Marie Bsumek, David Kinkela, and Mark Atwood Lawrence, eds., *Nation-States and the Global Environment: New Approaches to International Environmental History* (New York: Oxford University Press, 2013).

7. For an interesting chapter on the role of "ecological networks" in invasion biology, see, for example, Cang Hui and David M. Richardson, *Invasion Dynamics* (Oxford: Oxford University Press, 2017), 265–93.

8. Bruno Latour, *Reassembling the Social: An Introduction to Actor-Network-Theory* (Oxford: Oxford University Press, 2007).

9. For standard texts on animal studies, see, for example, Chris Philo and Chris Wilbert, eds., *Animal Spaces, Beastly Places: New Geographies of Human-Animal Relations* (London: Routledge, 2000); Giorgio Agamben, *Das Offene: Der Mensch und das Tier* (Frankfurt/Main: Suhrkamp, 2003); for the wealth of new research on human-animal relations on the German side, see, for example, Sven Wirth et al., eds., *Das Handeln der Tiere: Tierliche Agency im Fokus der Human-Animal Studies* (Bielefeld, Germany: transcript Verlag, 2015); Forschungsschwerpunkt Tier Mensch Gesellschaft, ed., *Den Fährten folgen: Methoden interdisziplinärer Tierforschung* (Bielefeld, Germany: transcript Verlag 2016); Forschungsschwerpunkt Tier Mensch Gesellschaft, ed., *Vielfältig verflochten: Interdisziplinäre Beiträge zur Tier-Mensch-Relationalität* (Bielefeld, Germany: transcript Verlag, 2017); for animal history, see Harriet Ritvo, *The Animal Estate: The English and Other Creatures in the Victorian Age* (Cambridge, MA: Harvard University Press, 1987); Ritvo, *Noble Cows, Hybrid Zebras: Essays on Animals and History* (Charlottesville: University of Virginia Press, 2010).

10. For the colonial forest service, see Gregory A. Barton, *Empire Forestry and the Origins of Environmentalism* (Cambridge: Cambridge University Press, 2002); S. Ravi Rajan, *Modernizing Nature: Forestry and Imperial Eco-Development 1800–1950* (Oxford: Clarendon Press, 2006); Brett M. Bennett, *Plantations and Protected Areas: A Global His-*

tory of Forest Management (Cambridge, MA: MIT Press, 2015); for the botanical gardens, see Donal P. McCracken, *Gardens of Empire: Botanical Institutions of the Victorian British Empire* (London: Leicester University Press, 1997); Richard Drayton, *Nature's Government: Science, Imperial Britain, and the "Improvement" of the World* (New Haven, CT: Yale University Press, 2000); for the zoological gardens, see Takashi Ito, *London Zoo and the Victorians, 1828–1859* (London: The Boydell Press, 2014); for the acclimatization societies, see Christopher Lever, *They Dined on Eland: The Story of the Acclimatisation Societies* (London: Quiller Press, 1992).

11. Michael Osborne, "Acclimatizing the World: A History of the Paradigmatic Colonial Science," *Osiris* 15 (2000): 135–51.

12. For recent research on the cultural encounters and "multiple hierarchies" in the context of "European" research expeditions in the nineteenth century, see, for example, Moritz von Brescius, "Humboldt'scher Forscherdrang und britische Kolonialinteressen: Die Indien- und Hochasien-Reise der Brüder Schlagintweit (1854–1858)," in *Über den Himalaya: Die Expedition der Brüder Schlagintweit nach Indien und Zentralasien 1854 bis 1858*, edited by Moritz von Brescius et al. (Cologne: Böhlau Verlag, 2015), 31–88.

13. Such as, for example, Heather Ellis, *Masculinity and Science in Britain, 1831–1918* (Basingstoke, UK: Palgrave MacMillan, 2017).

14. See, for example, Jürgen Osterhammel, *Die Verwandlung der Welt: Eine Geschichte des 19. Jahrhunderts* (Munich: CH Beck, 2009), 84–128; Kenneth Pomeranz, "Teleology, Discontinuity and World History: Periodization and Some Creation Myths of Modernity," *Asian Review of World Histories* 1, no. 2 (2013): 189–226; William A. Green, "Periodization in European and World History," *Journal of World History* 3, no. 1 (1992): 13–53; Robert I. Moore, "A Global Middle Ages?," in *The Prospect of Global History*, edited by James Belich et al. (Oxford: Oxford University Press, 2016), 80–92.

15. Lynn Hunt, *Measuring Time, Making History* (Budapest: Central European University Press, 2008); Giordano Nanni, *The Colonisation of Time: Ritual, Routine and Resistance in the British Empire* (Manchester, UK: Manchester University Press, 2012); Vanessa Ogle, *The Global Transformation of Time, 1870–1950* (Cambridge, MA: Harvard University Press, 2015).

PART I | The Nation State and the Unpredictability of Nature

The Transformation of an Ecological Policy

Acclimatization of Cuban Tobacco Varieties and Public Scandalization in the French Empire, c. 1860–1880

ALEXANDER VAN WICKEREN

The history of acclimatization is as much a story of failure as of success. Scientists, state officials, and amateurs all tried to cultivate foreign species in new environments during the second half of the nineteenth century, the great age of acclimatization. European states and their colonies went to great lengths to mobilize plants and animals for "improving" nature. Many histories have been written about the mobility of species, their impacts, and cultural receptions. But far fewer have studied failure, and especially the scandalization that came when states or scientists failed to deliver what they had promised.

One of these failures became a scandal in the expanding and liberalizing public sphere at the beginning of the Third Republic of France (1870–1940). After promising success, France's state tobacco monopoly administration was forced to admit publicly in 1875 that its effort to acclimatize tobacco from Cuba had failed. The administration had only a little over a decade previously been taken over by engineers trained at the *École Polytechnique*. The administration ramped up efforts to produce domestic Cuban tobacco. Engineers faced criticism for a lack of expertise from financial officials from the *Contributions indirectes* and their allied politicians, a significant blow to their professional prestige and a threat to their relatively newfound authority in the administration. In response, the French parliament, the *Assemblée nationale*, formed a publicized commission of inquiry.[1] Eugène Rolland, the general director of the state tobacco monopoly, was questioned at it. In front of the Assemblée, he admitted to failing in efforts to produce Havana-like quality tobacco with Cuban seed, despite earlier claiming the likelihood of success. The affair was highly embarrassing for Rolland, a decorated state official, not least because the acclimatization of tobacco was imagined to be a reasonably simple task.

Media, democracy, and political scandal became interconnected phenomena in France, which influenced the design and publicity of acclimatization efforts by engineers working for the administration. In the eyes of many

European observers, the French Third Republic was awash in scandal. As the German historian Frank Bösch has demonstrated, the last decades of the nineteenth century were increasingly shaped and unsettled by public scandals.[2] The growth of scandal reflected an expanding and liberalizing world of print. Daily newspapers, pamphlets, and books increased the public visibility of politicians and made it easy to criticize them. Print media brought to light sensitive issues, such as rumors and actual wrongdoings. At the same time, European parliaments gained greater political power, although their influence varied from state to state.

This chapter shows how state efforts to acclimatize tobacco in the French Empire became caught up in a public scandal in the late 1860s and early 1870s. The chapter uses the lens of the public sphere and scandal to connect European social history with the wider history of acclimatization and biotic globalization. Acclimatization had been a feature of European imperial expansion from its early origins in the Mediterranean and Atlantic islands well before 1492.[3] Acclimatization reached a first peak of interest in the 1860s to 1880s, an intensive period of introductions facilitated by a growing international network of botanical and zoological gardens, state scientific agriculture and forestry departments, and a lively interest by wealthy amateurs. The 1860s–1880s represented a key turning point in attitudes toward acclimatization. Extensive failures, such as failed tree planting efforts in South Australia and South Africa, led to increased scrutiny by the state and public.[4] In France, efforts to acclimatize gained popularity, reaching their pinnacle with the failed effort to transplant eland antelope from Southern Africa to France.

Efforts to cultivate Cuban tobacco occurred during this period of intense interest and growing dissatisfaction. French engineers had been focused on the centralization of tobacco reform in France by planting Cuban tobacco seeds since the 1860s.[5] Like many optimists, these engineers believed initially that it would not be difficult to cultivate Cuban seed tobacco. In the 1870s, these engineers began to focus more on the French colonies as better sites for certain crops. This shift aligned with the continuing failure to cultivate French tobacco, despite representations from engineers that the task could still be achieved. Public agitators and politicians began to question whether France could support quality tobacco. Debates about tobacco acclimatization had added potency because of tensions over the direction of France's tobacco administration by engineers, who had directed the tobacco administration since 1860. The engineers running the administration worried that criticisms undermined their authority and threatened to marginalize them within the state and scientific community. In the end, they decided to concede their mistakes and redirect their efforts to keep Cuban-like tobacco production alive. The Parisian ecological tobacco policy gradually shifted from France to its colonial territories.

The study of tobacco in France adds a needed domestic example to the larger imperial history of agriculture and environmental management in the French Empire. The historiography of nineteenth-century French agricultural and environmental reform has focused primarily on the French Empire. Michael A. Osborne[6] and Christophe Bonneuil,[7] among others, have argued convincingly that the new attention of French political and scientific elites to imperial acclimatization activities and agricultural policies emerged from the broader "civilizing mission" that underpinned much of nineteenth-century European imperialism.[8] The emphasis on the colonies has led scholars to overlook domestic examples that do not seem linked explicitly to the civilization mission ethos.

The study of acclimatization in domestic France can be understood using perspectives developed from European social and cultural history. German-speaking historians have produced a wealth of literature on the public sphere and public scandals in recent years.[9] The public sphere, which has been a central topic in the history of science roughly since the 1990s,[10] offers important insights into the scandal surrounding tobacco. Historians studying agricultural reform and acclimatization in the French Empire have so far neglected the literature on public sphere and scandal. This chapter aims to transfer the concept of public scandal and scandalization to research on tobacco acclimatization in nineteenth-century France. It argues that public scandals played an important role in the transformation of French national and imperial tobacco policy.

Historians of science understand public media spaces (including journals, exhibitions, and books) and eye-contact spheres (including stages, bars, and semipublic laboratories) as integral parts of knowledge production and expertise. Sociohistorical conceptualization of public scandal, however, helps us to see state experts and scientists as fragile actors who had to respond to the public sphere.[11] Historians convincingly argue that the interrelation of parliament spaces and an extension of print culture resulted in a new vulnerability of state actors[12]—a term, however, often also applied to scientists' public acting and staging.[13] Social historians have analyzed how state actors—such as monarchs, politicians, or administrative officials—became "victims" of scandalizing processes. Accusations first uttered in press and print often ended up in the national political spotlight.[14] What made the period so volatile was the expansion of the public sphere to include new groups such as journalists, members of political oppositions, and marginalized "peripheral" actors (for example, lower social classes).[15]

Finally, social historians have addressed the centrality of scandal by looking at social and cultural norms (mostly related to fields of sexuality and violence), which were objects of scandalized debate and concern.[16] By focusing on scandalized tobacco acclimatization, this chapter addresses a comparable issue in the field of science: the nineteenth-century idea of a "scientifically" governed state.[17] In this respect, a focus on scandal helps to understand the defense strategies of expanding scientific government administrations toward public accusation and indignation. Reactions by scientists in public scandals influenced the wider processes of professionalization happening within specific disciplines; in this instance, the developing field of polytechnic tobacco sciences. The chapter aims to show how actors scrutinized and challenged, yet also confirmed and consolidated, the intensifying connection between scientific experts and the modern state.

The chapter first outlines how French engineers drew on experiments throughout the French Empire to improve tobacco in the context of similar European projects in the mid-nineteenth century. Second, the chapter shows how public scandal became a feature of French public life in the late 1860s. The third part explores how the scandal turned into an issue of parliamentary debate, giving a voice to experiences of regional tobacco officials who were previously marginalized. Finally, it demonstrates how the scandal of failed tobacco experiments in France moved tobacco engineers to distance themselves from the national dimension of acclimatization and give more priority to an ecological policy focused mainly on the non-European territories of the French Empire.

Becoming Cuban

Cuban cigars gained popularity throughout the Atlantic world during the mid-nineteenth century as a result of the unique effects that they produced when smoked.[18] The popularity of Cuban cigars pushed growers outside of Cuba to try to grow a similar style of tobacco. Before 1860, Parisian state institutions showed little interest in acclimatizing tobacco. This changed in 1860, with the decision by the tobacco monopoly organization to purchase Cuban seed and introduce it via Paris to France. They hoped to improve the domestic production of raw tobacco for cigar wrappers and filler as a type of import substitution. Such initiatives were part of the general attempt by

France's state tobacco monopoly administration, which had been founded in 1810–11, to modernize the tobacco industry. The Régie, as it was called, was organized by a Parisian central office instructing tobacco officials in the *Départements*. The monopoly organization became responsible for the cultivation, processing, and trade of tobacco.

After 1860, officials of the tobacco monopoly organization began to select and instruct farmers to trial Cuban seed.[19] The administration advised farmers how to prepare experimentation grounds in their rural properties, how to sow the variety named *tabac de Havane*, and how to provide regular reports to the Parisian central monopoly administration. In Paris, tobacco engineers, *ingénieurs des tabacs*, had started ordering various regional Départements like Haut-Rhin and Alpes-Maritime[20] to undertake experiments with Cuban tobacco varieties. Engineers trained at the École Polytechnique, founded in 1794, had been systematically employed as technical staff in various French state administrations since the French Revolution. Polytechnic engineers, however, had only begun to form a technical "corps" in the administration after 1830, when they became systematically responsible for the modernization of French state tobacco manufacturing.[21] Once they gained more power in the central tobacco office after 1860, the engineers began to dream of harvesting tobacco of Cuban quality in France for Parisian manufacturers producing Cuban-style cigars for the domestic market. They equally sought to encourage the cultivation of tobacco in the empire, but they hoped to be able to satisfy domestic demand by French production.

Before 1860, the monopoly organization had been led primarily by officials from the Contributions indirectes, a tax collecting agency associated with the Ministry of Finance, whose focus on finances and trade had left the office ignorant of the active agronomic involvement of the central tobacco administration. In 1859, the Parisian director of Finances reported to the Ministry of Algeria and the Colonies "that today the tobacco administration became aware of the interest to enrich [*doter*] our overseas establishments with a cultivation of the richness equal to Havana, Porto-Rico, Brazil etc."[22] Yet, the tobacco officials did not see themselves as the active part in the improvement: "The Régie, as it understands itself, is nothing more than a trader [*marchand*] that buys from the producers without occupying itself with the conditions of cultivation and the secrets of preparation."[23] Although it would be right, as the director argued, that the Régie employed agents for the cultivation of tobacco in France, those officials were only hired for fiscal goals and to count the tobacco leaves of each plantation.

Internal changes around 1860 led the administration to pursue a more active scientific role in the production of tobacco in France and in its empire. The administration had no official position within the French colonies, but its engineers tried to encourage colonial planters by providing technical manuals. In the *Revue maritime et coloniale* from 1864, a compiled *Note* had been published explicitly for farmers in Madagascar, an island that many thought would support quality tobacco for Parisian cigar manufacturers. The *Note* emphasized the "essential importance" of seeds of "Havanese provenance" and especially those from the Vuelta Abajo, a Cuban tobacco cultivation valley that gained global fame in the mid-nineteenth century.[24]

In contrast to the acclimatization experiments in France, colonial tobacco resources were not exploited and improved under the supervision of the administration, but by private farmers and agricultural officials in Madagascar.[25] In the 1860s, French engineers supported these settlers, merchants, and farmers but did not see their efforts as being equal in importance or success to domestic efforts. This was not yet the state-funded imperial *mise en valeur* of the later nineteenth century and the interwar period,[26] but represented a more decentralized, "benevolent and non-exclusive patronage" of men on the spot envisioning the improvement of colonial tobacco in accordance with contemporary theories of free trade.[27]

In spite of the differences between metropole and colony, all the reform initiatives were part of a circulation of knowledge connecting the Spanish colony of Cuba, Paris, and the French colonies.[28] Parisian engineers had traveled to Cuba since the 1840s. The tobacco administration installed French experts on the island in 1862 to learn about local conditions. French engineers met local savants, entrepreneurs, and farmers, as well as slaves, to learn about tobacco. At the same time, Cuban tobacco specialists frequently visited Paris where they exchanged information with monopoly officials. Jean-Jacques Théophile Schlœsing, the director of the *École impériale d'application du service des tabacs*, drew on "practical" as well as "scientific" advice from Cuba to design the first experiments with *tabac de Havane* in Paris-Boulogne.

These experiments reflected the growing interest in acclimatizing foreign species in Europe and its colonies, a trend that had its scientific origins in the mid-eighteenth century. Mid-nineteenth-century France saw an intense scientific discussion led by Isidore Geoffroy Saint-Hilaire and the *Société d'acclimatation*, which downplayed the importance of natural climatic habitats and played a pioneering role in the international acclimatization movement. Saint-Hilaire's so-called transformist theory offered an alternative to further accounts on acclimatization, whether pessimistic or optimistic. The transfer of animals and plants, in Saint-Hilaire's view, was made possible by an intrinsic "adaptive potential" that corresponded to various types of environments. Saint-Hilaire's abstract claims on the "limited variability of type" served as a frame for numerous practical experiments with exotic flora. The French notion of "malleability" of species explained the wide range of exotic crops that were used in acclimatization.[29] Far from being similarly theoretical, the dictionary of the Académie des Sciences in 1835 defined the verb *acclimater* as "to accustom to the temperature and influence of a new climate."[30]

Finally, the transfer of expertise and tobacco seeds from Cuba to France and its colonies was part of a global agronomic and industrial movement. From the 1850s and 1860s onward, traveling scientists, merchants, and workers contributed to the transfer of seed material and agricultural and industrial knowledge from Cuban tobacco farms, workshops, and manufacturers to tobacco cultivation sites in the Dutch colonies of Indonesia or the U.S. state of Connecticut.[31] The British tried to grow Cuban seed in India.[32] In 1846, the Dutch consul Guillaume Lobé claimed that it would be possible to find "proper rules" enabling tobacco farmers to grow "in the north of Europe a tobacco quality close to the one of the island Cuba."[33] In port cities such as Bremen and Hamburg, where the production of cigars with Cuban-style labels had increased from the early nineteenth century, Hanseatic merchants (for example, F. H. Meyer) were important intermediaries for the transmission of industrial knowledge about tobacco from Cuba.[34]

Nationalizing Acclimatization

In 1868, as a result of the experiments in his garden in Paris-Boulogne, the tobacco administration's leading chemist, Schlœsing, touted the benefits that would come when France acclimatized Cuban tobacco in his publication *Le Tabac: Sa Culture au point de vue du meilleur rendement*.[35] In the same year, Louis Grandeau, an agricultural chemist who closely cooperated with Schlœsing, portrayed the acclimatization projects in an even more enthusiastic manner. An article in the *Journal d'Agriculture pratique* claimed: "He [Schlœsing] finally exposed the results for the cultivation experiments of tobacco imported from Havana into French soil [*sol français*] and showed that it is possible to clearly improve [*d'améliorer*] the quality of French tobacco [*tabacs indigènes*] by the frequently repeated introduction of seeds from Havana."[36] The very idea of a *sol français* as a frame with which to validate the results of the acclimatization of Cuban seeds echoed the paradigm of national applicability of knowledge, which had become more and more important in the course of the nineteenth century.[37]

Soon after, a group of publicly engaged representatives of the Contributions indirectes and their allied French politicians challenged the story that French engineers acclimatized Cuban tobacco. Beginning in 1869, just one year after Schlœsing's results were published, the agitators had published brochures, books, and petitions that discussed the difficulties associated with *Havane*'s introduction, among other problems with the tobacco administration. In 1871, an anonymous pamphlet *Au Gouvernement et à l'Assemblée Nationale* blamed the engineers for shortcomings in high-quality cigar production in France, dwelling on Schlœsing's "Lilliputian experience," which had misled the officials to entrust him with "exotic species."[38] Baron Charles-Alfred de Janzé, elected as deputy of the Département du Côtes-du-Nord for the French parliament, had made similar accusations two years before in *Les finances et le monopole du tabac*; a text that had been published as a segmented series in the *Journal de Paris* in 1868.[39] Published immediately after Janzé's account in 1869, Louis Koch's *De l'introduction des élèves de l'école polytechnique dans les manufactures de l'État et de ses conséquences* presented the harshest judgment on acclimatization, encompassing not only Schlœsing's garden, but also evidence from other parts of France:

> This professor, Mr. Schlœsing, much concerned himself with the acclimatization of exotic seeds in France. In this context he tried to establish a cultivation in miniature [*culture en miniature*], by planting the seeds in pots and boxes with mixed soil. . . . Due to his carefulness and precaution, the chemist was able to artificially produce plants of an exotic origin that had conserved the structure and the form of their provenance and also a large part of their original essence. However, planted outside his garden [*en plaine*], the varieties did not acclimatize without losing their former quality [*qualité primitive*]; it is known from thirty years of experience that all seeds that have been transplanted adapt to the nature [*prend la nature*] of the new soil and climate, which enables them to grow.[40]

The tone of the pamphlets directed toward a state institution was rough and clearly took advantage of the transformations of the liberal period of Napoleon III's Second Empire, an authoritarian regime established in France after 1852. While books or reports in regional journals or newspapers would have been seriously endangered by state censorship in the early 1860s, state restrictions on public assemblies, press, and other print media had been noticeably liberalized after 1869. Consequently, the numbers of daily newspapers rose, and they delivered daily attacks on the government in "violent terms."[41]

The pamphlets' criticism of the administration was no liberal affront against the government or a challenge to the existence of the state monopoly, but rather it served as a pretext to destabilize the monopoly's polytechnic-trained staff. All the pamphlets agreed that the overly scientific *ingénieurs des tabacs* who had studied at the École Polytechnique were responsible for the failed experiments with Cuban tobacco varieties. They pointed out that the curriculum at the École concentrated on theoretical subjects, such as mathematics and physics. When Koch vividly mocked the engineer's "insider relationship" (*coterie*), he emphasized the fact that "mathematical chemists," represented by Rolland and Schlœsing, had been responsible for the acclimatization's failure.[42]

Koch's words were not principally directed against mathematics or new scientific disciplines, such as agricultural chemistry, which had increased its public visibility since the 1840s.[43] By emphasizing the responsibility of state scientists, the pamphlets repeated stereotypes about the alumni of the elitist École Polytechnique that had circulated since the revolution of 1848. Since then, a growing number of French scientists had stood up against Parisian educational institutions. They caricatured the students of the École Polytechnique as "inept in the real world" and poked fun at the "cerebral otherworldliness" of their abstract mathematical knowledge.[44]

Instead of polytechnic engineers, the pamphlets generally called for financial officials from the Contributions indirectes to staff the tobacco monopoly administration. The authors of the pamphlets came from this marginalized group. They used the opportunity to try to reclaim authority after the engineers of the École Polytechnique had taken over after 1860. In one of the pamphlets, the tobacco office was described as "our service"[45] and the author, apparently a financial official, claimed to speak for a group of "commissioned staff" (*employés*) of "very numerous members of that administrative family."[46] Koch even explicitly revealed himself as an "Employé Superieur de la Direction générale des Manufactures de l'État,"[47] and the hostile wording of his text made sure that he was not to be confused with the administration's engineers carrying the same title. As the financial officials' emphasis on different groups competing for state functions showed, the scandalization of the tobacco engineers who supposedly failed to acclimatize Cuban tobacco was part of the battle for positions and competence in the French state between different administrative groups during the nineteenth century.[48] Certain politicians represented the criticisms of financial officers. They combined criticism against acclimatization experiments, the expertise of the officer, and ideas of free political speech in their texts. Baron Charles-Alfred de Janzé, for example, whose book had had a main influence on the financial officials,[49] exploited the acclimatization issue to position himself in the changing political landscape of the late Second Empire.[50]

Tobacco officials from the Contributions indirectes felt unjustifiably stripped of their traditional competence in state office and lobbied for a dismissal of the tobacco engineers of the École Polytechnique:

> The engineer must disappear in the interests of the financial administration, tobacco farmers, and consumers, in one way or the other, whether by his reunion with another administration, or by a healthier [*plus saine*] and normal reorganization of the tobacco service. Subordinated under the direction of an upright, intelligent, and enlightened [*éclairé*] administrator, chosen from the former descendants of the *École Polytechnique*, the interests of the state, the solid law, and finally also the interests of all would be guaranteed seriously and objectively [*appui sérieux et impartial*].[51]

Public calls for a more "healthy" and more "normal" administration might not have been routine in addressing governmental institutions in the press, but lines such as those cited above were routinely uttered in the later nineteenth century when agitators publicly attempted to destroy faith in politicians and high administration.[52] The officials' wording shows that problems of the acclimatization experiments had essentially been transformed into an argument for a subordination of the polytechnicians under the financial experts of the administration.

The tobacco administration's engineers reacted with a hectically compiled public defense campaign that tried to silence the accusations that had circulated since 1869. A few weeks after Janzé's book had been published, Eugène Rolland responded with an essay suggestively entitled *Réfutation de M. le Baron de Janzé intitulé les finances & le monopole du tabac*. For Rolland, the "extra-parliamentary form" of Janzé's pamphlet with its "increasingly personal" attacks threatened to leave a "disturbing [*facheuse*] impression on the readers who were less familiar with these sorts of questions." Rolland believed that "the scandal provoked with such insistence" could only be stopped when it "fell back [*retombe*] on its authors."[53] Stigmatizing writers like Janzé as provocateurs, the director of the central tobacco administration tried to ignore the accusations, asserting that the acclimatization experiments had failed, while trying to defend his institution in other instances.

Scientists in Parliament

The controversy in the late 1860s challenged the legitimacy of science in the administration, but it did not produce any change in policy or leadership. Parisian engineers continued to coordinate experiments domestically. The public scandalization of 1869–70s, and the public interest it generated, paved the way for a parliamentary inquiry that was set up in the political climate of the Third Republic established in 1871. The inquiry helped to generate a political stage for the revival of and heightened attention to the controversial acclimatization policy.

In 1873, an *Enquête parlamentaire*, headed by Victor Hamille, deputy of the Assemblée nationale, was commissioned to monitor the tobacco administration and determine possible paths for its reform. The state and the consumer, the inquiry argued, had a right to know if it would be necessary to limit or to extend the budget of the office and whether the staff of the monopoly organization should be reduced or enlarged. Among other issues, the parliamentary inquiry investigated the agricultural improvement attempts of the central tobacco administration and carefully analyzed if and how field trials for the improvement had taken place.[54]

The inquiry had been demanded by both scandalizer and scandalized. In July 1868, Janzé had presented his claims personally to the *corps législatif*, hoping that Napoleon III's constitutional reforms would enable the deputies to reorganize the tobacco administration.[55] Although his position received some applause among representatives in the Assemblée, the full demand was not considered worthy to be implemented.[56] Simultaneously, however, the administration's engineers had demanded that parliament adjourn the expected "reorganization of our service" until an inquiry had been established where engineers could explain themselves.[57] Both the destabilization of the *corps des ingénieurs* and the vision of legitimizing the engineers' standing in the new republic regime had been driving forces behind the scenes.

It was not only the actions of the Parisian central administration, financial officials, and polytechnic engineers, but also the peripheral experience with tobacco acclimatization in many French regions that made the inquiry a scandal. The parliamentary questionnaires toured the Départements. Yet some of the department's general directors in the Contributions indirectes, as well as a few Conseils généraux, emphasized their loyalty to the central administration by submitting rather ambivalent or even positive testimonies to the *tabac de Havane* inquiry.[58] Disregarding such views, most of the other regional interviewees regarded the experiments with Havana tobacco as unsuccessful. Following the Conseils généraux of departments like Haute-Saône and Lozère, "fine cigars," entirely made from Cuban tobacco varieties cultivated in France and manufactured in Parisian factories, had not improved since engineers had been in charge.[59] Other interviewees even complained of an acclimatization "without useful results"[60] and in Haute-Pyrénées, officials were aware of a clear tendency toward the variety's "degeneration."[61] Although the rather dense and compressed form of the questionnaire only revealed a general concern, many tobacco administrations in France's Départements seemed to have been slightly opposed to the introduction of Cuban varieties before the inquiry. The South Alsatian Département Haut-Rhin, which had not been represented in the inquiry because of the Prussian occupation of Alsace-Lorraine after 1871, nonetheless provided an important example for the perception of the variety and the Parisian administration. Before the occupation, Alsatians had occasionally delivered positive reports to the central administration, while the majority nonetheless rather depicted images of disease and infection. While Schlœsing, as mentioned above, had downplayed instances of "rust" (*rouille*) in his treatise from 1869, the officials had already noticed bigger spots of "rust" in the early 1860s. Although rust was not perceived as a constant problem but as a seasonal phenomenon, debates on treatment and countermeasures soon began in Colmar's tobacco office.[62] For mid-nineteenth century scientists, merchants, and tobacco officials, "rust" showed the limits to intensifying crop production on a global scale. Coffee plantations, in particular, had been frequently threatened by rust. As a consequence, botanists did more research on the phenomenon.[63] In the 1860s, regional tobacco officials in French Départements believed that Havana tobacco varieties were less suited to French environments. The issue of financial compensation for farmers in case of plague spreading in the experimental fields, in particular, resulted in accusations that the Parisian engineers were responsible for these calamities: "The administration confirmed to me that the farmers must count on its equity [*équité*], but not to an exaggerated favor that is inadmissible. The administration cares about information I have provided on the farmers' attempts concerning rust that could haunt [*envahir*] their plantations, but I do not doubt its generous intentions to guarantee the farmers' interest facing such an eventuality."[64]

Even a financial guarantee by the state in cases of acclimatization resulting in crop failure did not convince farmers in Haut-Rhin. In the eyes of some officials, the polytechnic improvement policy had failed completely. In February 1864, the director of Colmar's office concluded that it would be impossible to fulfil the expectations that were connected to the experiments with

Havana tobacco.[65] He expressed a long-standing concern with the Parisian policy, but these concerns were ignored in Paris.

Although the engineers cultivated the public image of successful acclimatization, doubts slowly began to take shape in the internal sphere of the tobacco administration. As early as 1861, Rolland confirmed to officials from Haut-Rhin that the success of the experiments with Havana seeds was limited. Nonetheless, he decided to restart the experiments by ordering provincial administrators to stick closer to instructions on Cuban-style agricultural techniques that appeared to "guarantee success."[66] Though criticism from departmental officials was not able to alter the engineers' view on the possibilities of improving tobacco in France, it was not taboo for the officials to address problems of the introduction of Cuban tobacco varieties—as long as they remained in the inner administrative circles of debate and did not bring it to public attention.

Toward the Colonies

The parliamentary inquiry of 1873 offered the first occasion for officials from the departmental "periphery" to publicly express doubt about the Parisian-led improvement policy. The inquiry also produced a statewide tableau revealing the results of French tobacco seed trials. The scandal further escalated in the wake of the inquiry. Threatened by decisions of a legislative organ with growing political competence, the Parisian engineers began to distance themselves from efforts to acclimatize the Havana variety. In line with the increase in colonial agricultural improvement after the war of 1870–71 and the loss of Alsace-Lorraine, engineers focused on a policy that concentrated on the French overseas colonies. This policy secured the position of the engineers in the monopoly administration.

Engineers felt rightly threatened. The commissioners of the parliamentary inquiry had threatened the engineers frequently by emphasizing the fact that tobacco engineers, as graduates of École Polytechnique, clearly received higher salaries than the lower administrators.[67] Rolland protested in an interview conducted by members of the inquiry's commission that "the tobacco administration and its director . . . had been the object of violent attacks."[68] Engineers were victims of their own hyperbole. Hoping to secure the tobacco administration's organization and the dominant influence of the *corps des ingénieurs*, Grandeau or Schlœsing had held up images of success that paved the way for a critical alienation from the acclimatization experiments on French national state territory. For the first time, state engineers were confronted with the full amount of the "erosion of the power" that they had gradually experienced in the decades after the Napoleonic era.[69]

Facing such scenarios of possible removal from or subordination to the administration, the polytechnicians altered their position. Initially, as Rolland pointed out in the interview, the administration had hoped to grow tobacco plants in France that could be specially used as cigar cover leaf. He admitted, however, that experiments in the Départements of Ille-et-Vilaine and Lot-et-Garonne—the only tobacco regions Rolland mentioned explicitly—had suffered from failure.[70] He tried to shift the blame to farmers. Things had gone awry because farmers were bound to apply "old-fashioned methods" and ignored instructions on Cuban-style cultivation that Parisian officials had distributed.[71] The strategy of distancing from acclimatization experiments in France was successful insofar as Victor Hamille's final report of the parliamentary inquiry from 1875 left the tobacco engineers' position untouched. For the French parliament, the polytechnic tobacco administration still appeared as a symbol of progress and success: "The commission, after having deepened its investigations to all segments [*rouages*] of the Régie's administrative organism, must truthfully declare that no part has shown traces of abuse [*abus*] that had been initially signalized to the commission. Therefore, the commission likes to pay tribute to the spirits of order, economy, and progress that characterize the Administration des-Manufactures de l'État."[72]

"Order" and "progress," main concepts of the "modernity" that had symbolized and stabilized the polytechnic engineers at least since the late eighteenth century, remained leading terms in the perceptions of the engineers. Consequently, the commissioners decided to provide "necessary credits to increase the cadre of staff charged with the surveillance of the cultivation and improve the position of the service's staff."[73]

Rather than referring to the long tradition of polytechnic engineers in the French state administration, the engineers' confirmation by parliament was legitimated by the new colonial improvement model that Rolland presented to the commission.[74] Thereby, tobacco engineers linked their *corps* with broader agendas encouraging colonial agriculture after 1870.[75] They justified this through the idea of a *mise en valeur* or "civilizing mission." Following the director, "analogous attempts" to the experiments in France had already been entertained in some parts of the French Empire; as, for instance, in Corsica, Algeria, Guyana, and Réunion, where experiments with Cuban tobacco had been already performed. Though Rolland saw "a certain chance to renew . . . initiatives successfully," which had been undertaken in the French Départements before, he highlighted the difficulties of precisely predicting their future results for all of the colonial territories.[76]

At the same time, however, French parliamentarians appreciated the engineers' proposed progressive policy of focusing on the colonies because it offered a substitute plan for the Alsatian tobacco cultivation that had been lost as a resource for the state's cigar production after the Franco-Prussian War in 1870–71. While tobacco from Bas-Rhin had already been included since the Napoleonic era, Haut-Rhin had delivered *tabacs de Havane* for cigar cover leaves to the manufacturers of Cuban cigars in Paris-Reuilly only since 1864.[77] As Rolland explained to the commission of the inquiry, the loss of Alsace had led to several problematic shortages because the region's raw tobacco had provided the largest amount of French cigar tobacco processed in the monopoly's state factories.[78] The loss of Alsace, coupled with the images of colonial improvement policies that fostered support for colonial tobacco of Cuban quality, paved the way for new collaborations among state institutions and extended the state policy of distributing knowledge about tobacco cultivation to the colonies. In 1874, when the inquiry was still in progress, the *Commission mixte de l'exposition permanente des Colonies* published a *Note sur la culture du tabac* for the parliamentary inquiry, in which conditions for successful cultivation of Cuban-like tobacco in France's overseas colonies were suggested. Following the parameters set out by the administration engineers in 1869, the role of soil composition and the problematic effects of nitrogenous fertilizer, as well as the choice of varieties, were highlighted.[79] Finally, the *Note* emphasized the importance of seeds from Havana yet it also warned that they had "not been naturally produced in France" and that this would be a "serious obstacle for the propagation of the Havana variety" in the colonies.[80]

Conclusion

The scandal over acclimatizing Cuban tobacco in France closed the chapter in the circulation of Havana varieties between Cuba and France and opened another one: French officials shifted attention to the colonies, rather than only to France, as a site of tobacco cultivation. In the 1870s, improvement policies of the central tobacco administration changed as they increasingly focused on the colonies. While current research argues that the shift toward the colonies was the result of a "civilizing mission" to improve colonial agriculture, this chapter underlines the importance of the French public and the scandalizing processes scrutinizing the polytechnic engineers in the French state tobacco office.

The story of French Cuban cigar production and the related ecological interventions is one of failure that led to a scandal. Acclimatization experiments often failed or had unexpected results but not all became scandals or even warranted attention.[81] Introducing the analytical category of scandalization shows the significance of public spheres in making these failures visible and stopping programs from continuing that might have succeeded. The nineteenth-century production and circulation of ecological knowledge can hardly be understood without considering these interrelated public spheres. An extending, liberalizing print culture and strengthening parliamentary power in the French Third Republic provided new spaces where agitators could refer to nonhuman actors to destabilize the social structure of the human world. How the public sphere and scandal influenced other acclimatization efforts will hopefully be the subject of future research by other scholars.

Notes

I would like to thank Corinna Kühn, Lena Rüßing, Pascal Schillings, and Jakob Vogel for their advice.

1. On this particular parliamentary inquiry, see Christophe Charles, *A Social History of France in the 19th Century* (Oxford: Berg, 1994), 155–56.

2. Frank Bösch, *Öffentliche Geheimnisse: Skandale, Politik und Medien in Deutschland und Großbritannien, 1880–1914* (Munich: Oldenbourg, 2009).

3. Alfred W. Crosby, *Ecological Imperialism: The Biological Expansion of Europe, 900–1900*, 2nd ed. (Cambridge: Cambridge University Press, 2004).

4. Brett Bennett, *Plantations and Protected Areas: A Global History of Forest Management* (Cambridge, MA: MIT Press, 2015).

5. Alexander van Wickeren, "Territorializing Atlantic Knowledge: The French State Tobacco Monopoly and the Globalization of the Havana Cigar around Mid-19th Century," in *Transnational Cultures of Expertise: Knowledge and the Rise of the Modern State*, edited by Lothar Schilling and Jakob Vogel (Berlin: De Gruyter, forthcoming).

6. Michael A. Osborne, *Nature, the Exotic, and the Science of French Colonialism* (Bloomington: Indiana University Press, 1994).

7. Christophe Bonneuil, "Mettre en ordre et discipliner les tropiques: Les Sciences du Végétal dans l'Empire Français, 1870–1940" (PhD diss., Université Paris-Diderot, 1997).

8. See also Patrick Petitjean, "Science and the 'Civilizing Mission': France and the Colonial Enterprise," in *Science across the European Empires, 1800–1950*, edited by Benedikt Stuchtey (Oxford: Oxford University Press, 2005), 107–28.

9. Most recently Caspar Hirschi, *Skandalexperten, Expertenskandale: Zur Geschichte eines Gegenwartsproblems* (Berlin: Matthes & Seitz, 2018); for a more general account, see Frank Bösch, "Kampf um Normen: Skandale in historischer Perspektive," in *Skandale: Strukturen und Strategien öffentlicher Aufmerksamkeitserzeugung*, edited by Christer Petersen and Kristin Bulkow (Wiesbaden, Germany: VS-Verlag, 2011), 29–48.

10. Steven Shapin, "Science and the Public," in *Companion to the History of Modern Science*, edited by Robert C. Olby (London: Routledge, 1990), 990–1007; Agustí Nieto-Galan, *Science in the Public Sphere: A History of Lay Knowledge and Expertise* (London: Routledge, 2016); Joris Vandendriessche, Evert Peeters, and Kaat Wils, eds., "Introduction: Performing Expertise," in *Scientists' Expertise as Performance: Between State and Society, 1860–1960* (London: Pickering & Chatto, 2015), 1–13.

11. Bösch, "Kampf um Normen," 33.

12. Bösch, *Öffentliche Geheimnisse*, 469.

13. Sybilla Nikolow and Christina Wessely, "Öffentlichkeit als epistemologische und politische Ressource für die Genese umstrittener Wissenschaftskonzepte," in *Wissenschaft und Öffentlichkeit als Ressource füreinander: Studien zur Wissenschaftsgeschichte im 20. Jahrhundert*, edited by Sybilla Nikolow and Arne Schirrmacher (Frankfurt: Campus, 2007), 273–85, 274.

14. Bösch, "Kampf um Normen," 38.

15. Bösch, *Öffentliche Geheimnisse*, 476.

16. Bösch, "Kampf um Normen," 35; for an empirical case study, see Rebekka Habermas, "Lost in Translation: Transfer and Nontransfer in the Atakpame Colonial Scandal," *Journal of Modern History* 86 (2014): 47–80.

17. Vandendriessche, Peeters, and Wils, "Introduction: Performing Expertise"; van Wickeren, "Territorializing Atlantic Knowledge."

18. Jean Stubbs, "El Habano and the World It Has Shaped: Cuba, Connecticut, and Indonesia," *Cuban Studies* 41 (2010): 39–67.

19. van Wickeren, "Territorializing Atlantic Knowledge."

20. On Alpes-Maritime: Anonymous, "Tabacs," *Département des Alpes-Maritime, Conseil général, Session de 1865—Rapport du Préfet et annexes: Procès-Verbaux des délibérations* (1865), 174–75; for archival evidence on acclimatization in Haut-Rhin, see below.

21. On the history of the French tobacco administration, Muriel Eveno and Paul Smith, *Histoire des monopoles du tabac et des allumettes en France XIXe–XXe siècles: Guide du chercheur* (Paris: Altadis, 2003), 22–25; for a general account on the role of polytechnic engineers in the French state, Bruno Belhoste and Konstantinos Chatzis, "From Technical Corps to Technocratic Power: French State Engineers and Their Professional and Cultural Universe in the First Half of the Nineteenth Century," *History and Technology* 23 (Autumn 2007): 209–25.

22. Cited in Guillaume Capus, Fernand Leuillot, and Étienne Foëx, *Le Tabac: Rendement et prix de revient—fabrication—production—action physiologique—régimes fiscaux—usages*, vol. 3 (Paris: Société d'Éditions Géographiques, Maritimes et Coloniales, 1930), 77.

23. Capus et al., *Le Tabac*.

24. Anonymous, "Note de la Direction générale des tabacs sur la culture du tabac les soins qu'elle réclame et ceux à donner à ses produits," *Revue maritime et coloniale* 10 (1864): 545.

25. In Madagascar, private botanical research initiatives had been introduced in the mid-nineteenth century, whereas France only sent state officials until the end of the nineteenth century. William Beinart and Karen Middleton, "Plant Transfers in Historical Perspective: A Review Article," *Environment and History* 10 (Winter 2004): 14.

26. See the chapter by Idir Ouahes in this volume.

27. David Todd, "A French Imperial Meridian, 1814–1870," *Past and Present* 210 (February 2011): 179.

28. van Wickeren, "Territorializing Atlantic Knowledge."

29. Michael A. Osborne, "Acclimatizing the World: A History of the Paradigmatic Colonial Science," *Osiris* 15 (2000): 140.

30. Cited in Osborne, "Acclimatizing the World," 137.

31. Stubbs, "El Habano and the World."

32. Kathinka Sinha Kerkhoff, *Colonising Plants in Bihar (1760–1950): Tobacco betwixt Indigo and Sugarcane* (Gurgaon: Partridge India, 2014), 99–100.

33. Guillaume Lobé, *Mémoire sur la culture du tabac dans l'île de Cuba* (Cayenne, French Guiana: Imprimerie du Gouvernement, 1845), 1.

34. F. H. Meyer's publication on Cuban tobacco was dedicated to manufacturers in Bremen and Hamburg: F. H. Meyer, *Aus der Havanna—Erfahrungen und Ansichten über die Fabrikation der echten Cigarren: Nebst Mittheilungen über Tabacksbau und Tabackshandel sowie nützlichen Winken für Fabrikanten* (Bremen, Germany: Meyer, 1854).

35. Jean-Jacques Théophile Schlœsing and Louis Grandeau, *Le Tabac: Sa Culture au point de vue du meilleur rendement—Combustibilité des feuilles, richesse en nicotine, etc., etc.* (Paris: Librarie Agricole de la Maison Rustique, 1868), 102–5. Schlœsing, however, also spotted "some accidents of rust appearing," which he mentioned rather randomly.

36. Louis Grandeau, "Culture de Tabac: Recherches expérimentales de M. Th. Schlœsing," *Journal d'Agriculture pratique: Moniteur des comices, des propriétaires et des fermiers* 32 (1868): 66–67.

37. On the entanglement between science and national ideas in the nineteenth century, see Ralph Jessen and Jakob Vogel, eds., *Wissenschaft und Nation in der europäischen Geschichte* (Frankfurt: Campus, 2002).

38. Anonymous, *Au Gouvernement et à l'Assemblée nationale—Réformes et économies administratives—Service des Tabacs: Réfutation du mémoire présenté au gouvernement et à l'Assemblée nationale par des ingénieurs des Manufactures de l'État ancien élèves de l'École Polytechnique* (Paris: L. Dupont, 1871), 13.

39. Baron Charles-Alfred de Janzé, *Les finances et le monopole du tabac* (Paris: A. Sauton, 1869), 32.

40. L[ouis] Koch, *De l'introduction des élèves de l'école polytechnique dans les manufactures de l'état et de ses conséquences: Détails destinés à compléter le judicieux et très-véridique rapport de M. le Baron de Janzé, député des Côtes-du-Nord, présenter au corps législatif dans sa séance du 21 Juillet 1868* (Paris: Imprimerie de E. Bossan, 1869), 21.

41. Roger Price, *The French Second Empire: An Anatomy of Political Power* (New York: Cambridge University Press, 2001), 269.

42. Koch, *De l'introduction*, 1–2.

43. For more general information, see Nathalie Jas, *Au carrefour de la chimie et de l'agriculture: Les sciences agronomiques en France et en Allemagne, 1850–1914* (Paris: Éditions des archives contemporaines, 2001).

44. Robert Fox, *The Savant and the State: Science and Cultural Politics in Nineteenth-Century France* (Baltimore, MD: John Hopkins University Press, 2012), 11, 35–36.

45. Anonymous, *Au Gouvernement*, 7.

46. Anonymous, *Au Gouvernement*, 2.

47. Koch, *De l'introduction*, 28.

48. Lutz Raphael, *Recht und Ordnung: Herrschaft durch Verwaltung im 19. Jahrhundert* (Frankfurt: Fischer, 2000), 175–76.

49. Anonymous, *Au Gouvernement*, 5. For the anonymous author, Janzé's book was an "extremely noteworthy book."

50. On Janzé and his political agenda, see Anonymous, "Baron de Charles-Alfred de Janzé," in *Dictionnaire des parlementaires français: Depuis le 1er Mai 1789 jusqu'au 1er Mai 1889, FES-LAV*, edited by Adolphe Robert, Edgar Bourloton, and Gaston Cougny (Paris: Bourloton, 1891), 402–3.

51. Koch, *De l'introduction*, 27.

52. Bösch, *Öffentliche Geheimnisse*, 13, 472.

53. Eugène Rolland, *Réfutation de M. le Baron de Janzé intitulé les finances & le monopole du tabac* (Paris: Librairie administrative de Paul Dupont, 1869), 5–6.

54. Anonymous, "Extraits des réponses présentées par les conseils généraux," *Annales de l'Assemblée nationale: Compte-rendu in extenso des séances—Annexes—Du 18 Décembre 1875 au 8 Mars 1876* 44 (1876): 168.

55. On the constitutional reform, see Price, *Second Empire*, 396; Daniel Prejko, *Gegen Minister und Parlament: Der Conseil d'État im Gesetzgebungsverfahren des Zweiten Französischen Kaiserreichs (1852-1870)* (Frankfurt: Klostermann, 2012), 316.

56. Rolland, *Réfutation*, 4.

57. Anonymous, *Au Gouvernement*, 7.

58. Anonymous, "Extraits des réponses présentées par les conseils généraux," 331.

59. Anonymous, "Extraits des réponses présentées par les conseils généraux," 333.

60. Anonymous, "Extraits réponses par MM. les Directeurs des Manufactures d'État," *Annales de l'Assemblée nationale: Compte-rendu in extenso des séances—Annexes—Du 18 Décembre 1875 au 8 Mars 1876* 44 (1876): 380.

61. Anonymous, "Extraits réponses par MM. les Directeurs des Manufactures d'État," 380–81.

62. The Controller of Colmar to the Inspector of Colmar, Neuf-Brisach, 3 Sept. 1864, Série 4/P/209, Archives départementales du Haut-Rhin (ADHR), Colmar, France. See also, The Director of Haut-Rhin to the Inspector of Haut-Rhin, Vesoul, 1. Aug. 1863, Série 4/P/209, ADHR, Colmar, France.

63. Stuart McCook, "Global Rust Belt: Hemileia Vastatrix and the Ecological Integration of World Coffee Production since 1850," *Journal of Global History* 1 (Summer 2006): 182.

64. The Director of Colmar to the Inspector of Colmar, Vesoul, 18 Mar. 1864, Série: 4/P/209, ADHR, Colmar, France.

65. The Director of Haut-Rhin to the Inspector of Haut-Rhin, Vesoul, 25 Feb. 1864, Série: 4/P/209, ADHR, Colmar, France.

66. Eugène Rolland to the Director of Haut-Rhin, Paris, 26 Feb. 1861, Série: 4/P/208, ADHR, Colmar, France; see also Eugéne Rolland to the Director of Haut-Rhin, Paris, 23 Jan. 1862, Série: 4/P/208, ADHR, Colmar, France.

67. Victor Hamille, "Rapport fait au nom de la commission d'enquête sur l'exploitation du monopole des tabacs et des poudres, sur la fabrication des tabacs et

l'organisation administrative de la Régie," in *Enquête parlemantaire sur l'exploitation du monopole des tabacs et des poudres* (Paris: Imprimerie nationale, 1876), 9–10.

68. Commission d'Enquête sur les tabacs: Séance du Jeudi, 27 May 1873, Série: C//3082, Archives Nationales, Paris, France.

69. Fox, *The Savant and the State*, 19.

70. Anonymous, "Réponses présentées par M. le Directeur général des Manufactures de l'État après examen et avis de son conseil d'administration," in *Annales de l'Assemblée nationale: Compte-rendu in extenso des séances—Annexes—Du 18 Décembre 1875 au 8 Mars 1876* 44 (1876): 168.

71. Anonymous, "Réponses présentées par M. le Directeur général des Manufactures de l'État," 169.

72. Victor Hamille, "Rapport fait au nom de la commission d'enquête sur l'exploitation du monopole des tabacs et des poudres, sur la fabrication des tabacs et l'organisation administrative de la Régie," in *Assemblée nationale, Année 1875, Impressions: Projets de lois, propositions, rapports, etc., Numéros 3420 à 3488* 46 (1875): 163.

73. Anonymous, "Note sur les travaux de la commission d'enquête," in *Annales de l'Assemblée nationale: Compte-rendu in extenso des séances—Annexes—Du 18 Décembre 1875 au 8 Mars 1876* 44 (1876): 135.

74. Anonymous, "Réponses présentées par M. le Directeur général des manufactures de l'État," 174.

75. See Bonneuil, "Mettre en ordre et discipliner les tropiques," 23–162.

76. Bonneuil, "Mettre en ordre et discipliner les tropiques," 168.

77. The Director of Haut-Rhin to the Inspector of Haut-Rhin, Vesoul, 16 Aug. 1864, Série: 4/P/209, ADHR, Colmar, France; see also the memorandum of Eugène Rolland to the Directors of the Departments, Paris, 10 Feb. 1864, Série: 4/P/209, ADHR, Colmar, France.

78. Commission d'Enquête sur les tabacs: Séance du Jeudi, 27 May 1873.

79. Anonymous, "Note sur la culture du tabac, publiée par les soins de la Commission mixte des tabacs de l'exposition permanente des Colonies," in *Annales de l'Assemblée nationale: Compte-rendu in extenso des Séances—Annexes—Du 18 Décembre 1875 au 8 Mars 1876* 44 (1876): 452. An extended version was published in 1875: Exposition permanente des colonies: Comission mixte des tabac, Note sur la culture des tabacs, Paris, Sept. 1875, Série: C//3082, Archives Nationales, Paris, France.

80. Anonymous, "Note sur la culture du tabac," 454. The shift in the policy, however, was not complete but rather gradual. The loss of Alsace therefore encouraged engineers, as they claimed in the inquiry, to try new "extensions" of the cultivation of cigar tobacco in "certains Départements": Hamille, "Rapport fait au nom de la commission," 23.

81. Brett Bennett, "The El Dorado of Forestry: The Eucalyptus in India, South Africa, and Thailand, 1850–2000," *International Review of Social History* 55 (2010): 27–50.

Securing Resources for the Industries of Wilhelmine Germany

Tropical Agriculture and Phytopathology in Cameroon and Togo, 1884–1914

SAMUEL ELEAZAR WENDT

In a lecture at the German Colonial Congress in Berlin in 1905, the botanist Walter Busse warned that pest infestations threatened agricultural productivity in Europe's tropical colonies.[1] Busse had just returned from a so-called phytopathological expedition to Cameroon and Togo and gave an alarming view of Africa, where "local insects, living on rainforest or veld plants, have turned voraciously on some of the introduced agricultural crops more appealing to them, . . . and have proliferated on a frightening scale."[2] In some areas, Busse pointed out, the infestation was so significant that "the cultivation of certain crops could not be continued."[3] Busse stressed the importance of plant pathology for colonial agricultural production in the light of major crop infestations, especially in Africa.

Studying Busse and others, this chapter examines the transimperial transfer of plants and knowledge about their cultivation in the German colonial empire. In the age of high imperialism, Germany sought to overcome its dependency on imports of raw materials from Britain, France, and the Netherlands—competing colonial powers. The transfer of knowledge and biota in the German Empire was made possible through a network of industrialists, colonial institutions, botanical gardens, experimental stations, experts, and expeditions. It included both human and nonhuman actors.

German colonial agriculture drew on precedents from other European powers in Africa and Asia. Agronomists and botanists sought to expand their knowledge of tropical agriculture and phytopathology in order to be able to create viable agricultural plantations in the tropics. Plant pathogens, both native and exotic, threatened the prospect of securing raw material resources for the industries of Wilhelmine Germany from its colonial plantations. By examining the development of cotton, rubber, and cocoa plantations in the colonies of Cameroon and Togo, this chapter illustrates how colonial agriculture and phytopathology helped to transform, perceive, observe, and communicate environments and nonhuman agency.

Walter Busse and the German Network of Tropical Agronomy

On August 28, 1904, the steamer *Eleonore Woermann* arrived at the coast of Cameroon and pulled into the port of Victoria. On board was the German botanist, tropical agronomist, and expert in plant pathology, Walter Busse (1865–1933), assigned by the German Colonial Economic Committee (Kolonial-Wirtschaftliches Komitee, KWK) to locate, study, identify, and catalog the pathogens that had befallen many of the cash crops on the plantations in the West African colonies of Togo and Cameroon. His mission was to devise means against the moths, beetles, and fungi that had in some cases severely diminished the crops, notably rubber, cotton, and cocoa.[4]

Soon after the German Empire had acquired and legalized its colonial possessions in Africa in the 1880s and 1890s, an array of scientific expeditions and fortune-seekers came to investigate the territory.[5] They were interested in exploring the geography of the region, its peoples and customs, flora and fauna. Botanists hoped to find new cash crops that promised to yield great profit or to identify areas that could suitably support introductions.[6]

Two colonial societies, the German Colonial Society (Deutsche Kolonialgesellschaft, DKG) and the KWK, provided most of the funds for these expeditions. Both interest groups, founded in the years 1887 and 1896 respectively, had managed to recruit as members important individuals and corporations from German society, politics, industry, and commerce—such as Deutsche Bank, the industrialist Friedrich Alfred Krupp, and the merchant and ship owner Adolph Woermann.[7] The membership records of the DKG published in 1914 included 61 municipalities and 41 banking houses; approximately 150 business, industrial, scientific, and technical institutes; the chambers of commerce of most major German cities; a large number of colonial companies, and nine missionary societies.[8] Furthermore, many leading German scientists, businessmen, merchants, bankers, engineers, industrialists, and lawyers belonged to the board of directors of the KWK.[9] By incorporating the Colonial Economic Committee into the German Colonial Society as its economic committee, both groups merged in 1902.

The KWK sought explicitly to advance the economic and scientific development of the German colonies by focusing on the cultivation of useful tropical plants needed for industrial processing—primarily rubber, cotton, and cocoa.[10] To this effect, they sought to stimulate the establishment of planta-

TABLE 1 Value of German imports, in billions of marks, 1888–1907

Imported goods	*1888*	*1892*	*1896*	*1900*	*1902*	*1904*	*1906*	*1907*
Raw materials	1.49	1.66	1.89	2.8	2.56	3.18	4.03	4.43
Manufactured goods	0.88	0.85	0.94	1.15	1.1	1.22	1.67	1.88
Cattle, foods, luxury foods (tobacco, coffee, cocoa, sugar)	0.91	1.5	1.48	1.76	1.97	1.96	2.32	2.43

Source: Kolonial-Wirtschaftliches Komitee, ed., *Unsere Kolonialwirtschaft in ihrer Bedeutung für Industrie und Arbeiterschaft*, Beihefte zum Tropenpflanzer, no. XIII/3 (Berlin: E. S. Mittler & Sohn, 1909), 7.

TABLE 2 German imports, exports, and trade balance, in billions of marks, 1888–1907

	1888	*1892*	*1896*	*1900*	*1902*	*1904*	*1906*	*1907*
Imports	3.3	4	4.3	5.8	5.7	6.4	8	8.7
Exports	3.2	3	3.5	4.6	4.7	5.2	6.4	6.9
Balance	–0.1	–1	–0.8	–1.2	–1	–1.2	–1.6	–1.8

Source: Kolonial-Wirtschaftliches Komitee, ed., *Unsere Kolonialwirtschaft*, 6.

tions in the colonies. German supporters believed that tropical raw materials could be grown in the colonies in enough quantities to satisfy industrial demand and loosen Germany's dependency on other European colonies and countries. Germany could only entirely satisfy its domestic need for coal, salts, and ceramic raw materials. It had to rely on imports for the supply of almost all other raw materials. As shown in table 1, between 1888 and 1907 the value of imported raw materials almost tripled, from 1.49 billion to 4.43 billion Marks. Likewise, due to the decrease in agricultural productivity following industrialization, the amount of imported foodstuffs, luxury foods (coffee, sugar, tobacco, tea, and cocoa), and cattle doubled in the same period. As shown in table 2, the described development led to an increase in Germany's trade balance deficit, which in turn highlights the perceived need for and exigency of developing colonial agriculture and achieving supply autarky.

Following the example of other European powers, primarily Britain and the Netherlands, which developed institutions for economic botany and colonial tropical agriculture,[11] the KWK actively participated in the creation of a network of institutions, experimental stations, and qualified personnel for the production and circulation of knowledge, plant breeding, and the experimentation with and allocation of introduced plants. The KWK drew on the assistance of governmental institutions such as the Botanical Garden in Berlin (Botanischer Garten Berlin), the Botanical Museum and Botanical Laboratory for Merchandise Knowledge in Hamburg (Botanisches Museum und Botanisches Laboratorium für Warenkunde), the Botanical Research Center for the German Colonies in Berlin (Botanische Zentralstelle für die deutschen Kolonien), and the Research Gardens in the Colonies (Versuchsstationen in den Kolonien). The collaboration between the KWK and this network of state institutions and personnel facilitated the consignment of hundreds of living tropical plants and thousands of batches of seeds for cultivation in the colonies. It also participated in recruiting and training the personnel selected for colonial service. These students learned about the cultivation of tropical plants and colonial botany and plant pathology in the Botanical Garden in Berlin and in the German Colonial School (Deutsche Kolonialschule) in Witzenhausen. Additionally, the KWK was paramount in organizing and financing expeditions, and in transferring useful plants through inter- and transimperial networks.[12]

The German colonial venture was not led exclusively by the German state and government. Many associations, colonial pressure groups like the KWK, business houses like Godeffroy or Woermann, merchant networks, other nongovernmental organizations like the Pan-German League (Alldeutscher Verband), and even university scholars were actively engaged in shaping colonial policy and politics.[13] In the case of the KWK, the government relied upon their expertise and network, and expected assistance in solving agricultural difficulties.

The KWK assigned Walter Busse to conduct the first botanic-pathological expedition to Cameroon and Togo. After studying botany in Berlin and Freiburg, and obtaining his doctorate degree in 1892, he began working as a scientific assistant at the bacteriological department of the Imperial Board of Health in Berlin (Kaiserliches Gesundheitsamt), conducting research in the field of applied chemistry and food hygiene, specifically condiments. After obtaining his postdoctoral lecture qualification at the Friedrich-Wilhelm University of Berlin in the year 1900, he taught bacteriology and tropical agriculture there until 1903, when he left academia for continuing field research and subsequently a career in public office. His first botanical and agricultural expedition assigned by the KWK and at the request of the government of German East Africa (Kaiserliches Gouvernement in Deutsch-

Ostafrika), took him to that colony in 1900. There, he collected material and recorded information about agricultural crops and their diseases, as well as on medical and poisonous plants utilized by the local population.

After returning to Germany, the Imperial Colonial Office (Reichskolonialamt) awarded Busse the Buitenzorg-Scholarship in 1902, which gave young botanists the opportunity to undertake an educational and scientific journey to the tropics. At Buitenzorg, Busse continued his scientific studies on the cultivation of tropical cash crops and broadened his understanding of the treatment of various diseases affecting plants. The grant was named after the renowned botanical and experimental garden established by the Dutch in Java, which was considered by many, including leading German scientists, the model institution for conducting research in tropical botany and agriculture.[14] Florian Wagner's contribution to this volume shows how the botanical garden at Buitenzorg significantly contributed to shaping and reforming agricultural theory and practice since the 1900s.

Back in Germany in 1903, the administration of German East Africa once again appointed him to lead an expedition to the colony in order to study the techniques that native communities based in the Usambara Mountains applied to the cultivation of cereals. In 1904, the KWK nominated him to study the pathogens afflicting the rubber, cotton, and cocoa plantations in Cameroon and Togo because he was believed to be a leading expert in tropical agriculture as a result of his travels and studies.[15] His trip focused entirely on cash crop production issues.

Industrialization, Mass Consumption, and the Need for Tropical Raw Materials

In the course of the nineteenth century, rubber, cotton, and cocoa gained great importance as raw materials for industrial processing in Europe and the United States. Although cacao was known to Europeans since the discovery of the Maya and Aztec lands in the fifteenth and sixteenth centuries, its consumption in Europe remained a luxury accessible only to the socially and economically privileged classes well into the first half of the nineteenth century. During this time, production remained centered in Central and South America.[16] A significant shift in perspective, as well as technological achievements, prompted the mass consumption of chocolate. While cacao was long considered to have pharmaceutical properties and was thus advised as medicine, in the nineteenth century its nutritional value was advocated. Cacao could enhance the nourishment of the general population.[17] This shift in perception benefited from, and was supported by, the achievements of chocolatièrs and early food industrialists, including the invention of cocoa powder and the development of milk chocolate. These new products, made possible by the mechanization of production processes and advances in food technology, allowed for the manufacture of relatively inexpensive products, which facilitated access to and thus general consumption of chocolate.[18] By 1912, annual cocoa consumption in Germany had reached close to 0.82 kilograms per person.[19] Cacao had become a significant ingredient in the manufacture of cocoa products as well as an important foodstuff, and Germany became the second largest importer of cacao (approximately 54,000 tons per year) after the United States (57,000 tons), followed by Great Britain (28,100 tons).[20] Hence, it was an important concern for German food industrialists to lessen their dependency on foreign cacao imports by developing their own supply sources in the colonies.

In the case of cotton, we can trace a similar path. Here, too, the mechanization of production processes, cotton spinning, and weaving made cotton fabrics more accessible to broader consumer groups following the expansion of the textile industry, which began soon after the unification of Wilhelmine Germany in 1871. Prior to that, spinning and weaving were essentially cottage industries, processing mostly wool and linen. The expansion and concentration brought about by industrialization replaced these fabrics in favor of cotton and led to a tremendous increase in this branch of industry, tripling the number of looms between 1870 and 1910, from approximately 84,000 to about 260,000.[21] By 1909, Germany's cotton industry was the largest on the European continent and the third largest in the world. The value of its output was the most considerable of all domestic industries and constituted the nation's most important export product, making it a question of national interest to sustain this branch of industry, more so as it was almost entirely dependent on the import of raw cotton from overseas. In 1902, Germany imported approximately 500,000 tons of cotton.[22] Roughly 70 to 80 percent of the supplies were imported from the United States, while India and Egypt remained secondary suppliers.[23] This dependency on imports from the United States, especially after the cotton famine of the 1860s during the American Civil War, led textile industrialists to raise the *Baumwollfrage* (cotton question) and to promote cotton production in the colonies. One of these prominent industrialists was Karl Supf (1855–1915). He advocated raw material supply autarky through supplies produced in the colonial realm, for which he and the tropical botanist, Otto Warburg (1859–1938), founded the KWK in 1896.

Rubber also became an indispensable raw material during the nineteenth century. Even though Christopher Columbus and other conquistadors, missionaries, and settlers knew about the widespread uses of caoutchouc (*Castilloa elastica*) among the Aztec and Mayan peoples, for example to make cured rubber balls (*ōllamaloni*) used in the ritual ball game *ullamaliztli*,[24] Europeans did not pay much attention to this raw material until the French mathematician and explorer Charles Marie de La Condamine (1701–74) returned from his expedition to South America and published his *Relation abrégée d'un voyage fait dans l'intérieur de l'Amérique méridionale* in 1745. Here, he described the knowledge of and different ways in which rubber (*Hevea brasiliensis*) was used by the Mainas[25] of the Amazon basin, and introduced the term *caoutchouc*, which is thought to stem from the Quechua *kawchu* (vegetable juice) or *káuchuk* (tears of the tree),[26] as indicative of the described organic substance. La Condamine also brought back with him some samples, as well as items made from cured caoutchouc, such as waterproof shoes, syringes, and tubes.[27] By 1760, only fifteen years after La Condamine's return, manufacturers had embraced the new material and were impregnating boots and bags with it, utilizing its water-repellent properties. In 1768, the first tubing and catheters were produced for construction and medical purposes. Edward Nairne and Joseph Priestly, who were independently experimenting with caoutchouc and different solvents like turpentine, discovered the method for making erasers in 1770. In 1803, the first factory for the manufacture of rubber braces opened in France, while rubber-coated telegraph cable testing began in Germany in 1811.[28] Nonetheless, these early products manufactured out of uncured rubber did not handle temperature changes well, becoming brittle in the winter and sticky in the summer, which rendered them unfeasible. It was only after the discovery of the vulcanization process in the 1830s, by which rubber is cured through the addition of sulfur and the exertion of pressure and heat, that the manufacture of durable rubber goods was made possible.[29] As a result, the demand for this raw material grew enormously. Here too, Germany played a key role, as its rubber industry had become the third largest in the world by the 1890s. By 1910, the world's natural rubber production was about 90,000 tons, of which the German rubber industry processed approximately 14,000 tons.[30]

Therefore, it was a paramount concern for the German colonial lobby to foster the development of plantations in the colonies in order to satisfy the demand for these raw materials through yields produced in Germany's own tropical possessions. This strategy would alleviate the need to import goods from other colonies and secure Germany's economy, which became strongly protectionist in the 1880s. Even though Germany had almost become a free trade country in the 1870s, desires for a greater degree of economic independence and for the protection of strategic industries arose due to economic uncertainties that followed the depression of 1873.[31] The increasing competition for empires and external markets between European powers, as well as the rise of economic nationalism and militarism, also contributed to the implementation of ever higher protectionist barriers.[32]

Plant Transfers and the Development of Cash Crop Plantations in Cameroon and Togo

German colonial officials, agronomists, botanists, investors, and planters were convinced that the best way to produce tropical commodities was through cultivation in modern plantations. Central to their understanding was the belief that any tropical crop could be cultivated extensively, and its efficiency and output maximized.[33] Advancements in botany, chemistry, and agronomy promised to yield the tools necessary to cope with the challenges posed by unknown and difficult environments, such as the African rainforests or other so far "underused" lands. Expertise and capital procured by corporate enterprises would ensure the success of the undertaking, which had highly symbolic value; as, to quote Corey Ross, "plantations were much more than commercial enterprises; they were also incarnations of European agronomic knowledge and symbols of European power,"[34] reflecting the supremacy of European-based knowledge, technique, and rationality. Still, before Germans had established colonial rule over Cameroon and Togo, African societies were already cultivating certain cash crops like cotton or cacao. Their practices of smallholder production challenged German understanding of cash crop cultivation. Furthermore, nonhuman actors such as beetles, moths, and fungi soon began to have an impact on the course and shape of German colonial agro-business.

The Spanish and Portuguese had brought cacao tree seeds to their offshore island colonies in West Africa from South America, notably to São Tomé and Bioko (formerly known as Fernando Póo). From there, Africans transported them to the western continental coastal regions in the 1870s, where the local population began to cultivate the crop widely. During the nineteenth and twentieth centuries, this region turned into the most important area of cocoa cultivation, producing approximately three-quarters of annual global production.[35] In Togo, however, cultivation did not follow the path described above. Instead, cacao cultivation was initiated by German planters and colonial officials in 1900, using seeds and seedlings imported from Cameroon.[36] Around the same time, German officials introduced cacao varieties from Mexico and Venezuela to the region. These foreign varieties were to interbreed with local varieties in the botanical garden of Victoria, Cameroon, in order to improve the quality and taste of the cacao seeds and to increase their natural resilience to vermin and diseases.[37] These hybrids would later be transferred to the large plantations run by corporate enterprises and stock corporations like the Westafrikanische Pflanzungsgesellschaft "Victoria" AG (Berlin), the Westafrikanische Pflanzungsgesellschaft "Bibundi" AG (Hamburg), or the Moliwe Pflanzungsgesellschaft D.K.G. (Hamburg), situated around the western base of Mount Cameroon.[38]

Before German colonial rule, cotton had been grown and processed into yarn and cloth in Togo. It was strongly ingrained within local economic, social, and cultural practices that German colonial rule sought to break open in order to transform traditional cotton farming into an export-oriented agro-business. The Ewe traditionally grew cotton on fields mixed with other plants such as maize or yams, which not only provided them with food but moreover ensured their sustenance irrespective of fluctuating cotton economic cycles and prices. Nonetheless, the colonial government of Togo with assistance of the KWK employed four African American experts from Tuskegee Normal and Industrial Institute (Alabama) to instruct the Ewe how to plant and harvest cotton in a "rational" and "scientific" way.[39] After James Nathan Calloway (b. 1865), John Winfrey Robinson (1873–1909), Allen Lynn Burks, and Shepherd Lincoln Harris (d. 1902 in Togo) arrived in the colony in December 1900, they established a cotton farm near Tove.[40] Building on plantation models they knew from the United States, they chose to ignore the accumulated experience and agricultural and meteorological knowledge of the local population. When the rains began in July, all the different cotton varieties planted on the experimental fields—whether introduced from the United States, Egypt, Brazil, or Peru—started to rot.[41] However, this was not the only setback. A variety of different moths, beetles, and fungi started to affect the introduced cotton varieties dramatically. Meanwhile, the cotton traditionally grown by the Ewe remained largely unaffected. Additionally, with the transfer of cotton varieties launched by the Tuskegee experts, cotton rust (*Phakopsora gossypii*) was introduced into Togo, affecting all cotton varieties. As a measure to counteract these afflictions, planters created the Togo Sea-Island hybrid by interbreeding the local cotton with overseas varieties. They then distributed the seeds of this new variety to all farmers and also sought to match and improve the overall quality of the cotton grown, as the Ewe continued to grow cotton traditionally, interspersing their yam and corn fields with this crop.

In contrast, the available primary sources and secondary literature suggest that the local population had no use for local rubber-bearing plants, such as *Kickxia elastica* trees or the *Landolphia heudelotis* and *Landolphia owariensis* liana, before rubber industries were developed in Europe and the United States. In the coastal region of Victoria, near Mount Cameroon, German merchants created the first rubber plantations in the 1880s, while wild rubber (*Landolphia* liana) was increasingly gathered in the hinterland after the turn of the century.[42] In 1898, the German botanist Rudolf Schlechter (1872–1925) introduced *Kickxia* seeds from the British colony of Laos.[43] It was the KWK that had appointed Schlechter to lead the expedition, following a recommendation from the botanist Otto Warburg. Schlechter not only contributed to transferring rubber-bearing varieties from other colonial realms to the German colonies, he also established and oversaw the development of rubber plantations in Togo and Cameroon.[44]

Phytopathology and the Making of *Schädling* (Vermin)

Humans have for thousands of years tried to protect plants from harmful organisms and other damaging influences like hail or frost. In antiquity and the Middle Ages, magical-religious beliefs informed assumptions about plant disease, embedded in a holistic-cosmological approach. In this view, natural events like droughts or the devastation caused by insects were interpreted as manifestations of divine will.[45] Insects like locusts were regarded as harmful or noxious, while others—the vast majority—passed unnoticed with regard to human needs and the prevailing order of things.[46] The outbreak of rusts and other fungal diseases was attributed to the wrath of God, as he controlled the weather changes that were believed to bring about those diseases.[47]

A religious perspective remained predominant in Europe into the eighteenth century, when a new order emerged, driven by scientific and technological optimism. The invention of the microscope, generally attributed to the inventor Cornelius Debbel (1572–1633) and to the lens grinder Zacharias Janssen (1588–1631), proved to be pivotal for the emergence of plant pathology, as it allowed the identification and study of previously unknown organisms and structures.[48] In 1729, the Italian botanist Pier Antonio Micheli (1679–1737) described and illustrated various genera of fungi and their reproductive structures. He suggested that fungi develop from the spores they produce rather than through spontaneous generation, thus challenging the predominant model of his time, which took the microbes and spores as the result rather than the cause of a disease.[49] The idea that diseases could be caused by one organism growing within another seemed so unlikely that nobody accepted the experimental evidence Micheli obtained. Similarly, the French naturalist Benedict Prevost's (1755–1819) accurate conclusions on the origin of smut disease in wheat (contamination with smut spores) were neglected by the French Academy of Science.[50] Also, Miles Joseph Berkley's (1803–89) findings on the origin of potato blight, which he properly linked to a fungal infection, were neglected by contemporaries who were convinced that harmful environmental conditions were causing the disease.[51] The contributions of Louis Pasteur (1822–95) and Robert Koch (1843–1910), who unequivocally identified microorganisms as agents of disease in humans and animals, provided the basis for the germ theory of disease, which ultimately led to the end of the theory of spontaneous generation.[52] Amongst others, Heinrich Anton de Bary (1831–88), Thomas Jonathan Burrill (1839–1916), and Adolf Eduard Mayer (1843–1942) attested to these findings for different illnesses observed in plants, thereby laying the groundwork for the scientific study and subsequent abatement of plant diseases: phytopathology.[53]

While the first phase of phytopathology focused on the study and classification of fungous pathogens, other agents were discovered along the way. Following Koch's findings on the connection between bacterial infection and disease, Thomas Burrill discovered in 1882 that the fire blight disease of pear and apple trees was caused by bacteria. Soon after, the bacterial etiology of diseases afflicting potatoes, corn, cucurbit, sugar cane, and other plants was revealed.[54] A hitherto unknown pathogen was revealed in tobacco plants while studying the nature of yellow mosaic disease. Although the initial studies sought to find fungal or bacterial causation, Martinus Willem Beijerinck (1851–1931) finally concluded that microorganisms were not the cause of the disease. As the suspected agent could not be extracted through filtration and the probes also diffused to some degree in the Petri dish nutrient medium, he described the unknown agent as a *contagium vivum fluidum*, a contagious living fluid (i.e., a virus), as fluids were known to diffuse in agar.[55] The continuous search for the causation of plant diseases has led to the discovery of other types of pathogens: nematodes, protozoa, mollicutes, viroids, and, most recently, prions.[56]

As Sarah Jansen has shown in her seminal work on the emergence of the concept of *Schädling* (vermin) as a scientific and political construct in Germany, during the second half of the nineteenth and the first two decades of the twentieth century the relationship between humans and the natural sphere was severely reappraised and the catalog of harmful organisms significantly amplified. Those formerly disregarded or only considered to be noxious were now understood as harmful pests that needed to be contained or eliminated using the methods, tools, and chemical substances developed for pest control (*Schädlingsbekämpfung*).[57] As shown, different events and processes contributed to this shift. First, the introduction of the potato late blight and the potato beetle, as well as the vine pest and the vine fretter, led to devastating effects on crops in Western Europe and the United States. Second, new trends in scientific forestry, agronomy, and botany not only increased the interest in and exigency of promoting the systematic study of plant pathogens, but also promoted the aforementioned perceptional and cognitive shift.[58] In applying scientific methods for commercial ends, they aimed to achieve the most efficient means of producing timber or cash crops, thereby oversimplifying ecological complexities by excluding noncommercial species, nonhuman actors, and in some cases even human inhabitants as external to the production process.[59] Also, as Jansen and others have shown, the apprehension and categorization of nonhuman actors, and their *worth*, was now determined by concepts derived from economics—utility and loss—ultimately categorizing insects and other nonhuman entities as beneficial organisms or corrupting vermin.[60]

Walter Busse pointed to the correlation between corrupting agents, medicine, hygiene, and plant pathology in his lecture at the German Colonial Congress in Berlin in 1905, when he stated that "the same phenomenon that has already been observed for civilized man and domestic animals, is also becoming evident [in the field of plant pathology, SEW], expressed in a variety of different hygienic measures."[61] Furthermore, he specified the tasks of plant-hygiene (*Pflanzenhygiene*) as the abatement of diseases, preventing the proliferation and dispersal of pests within a certain country or geographical region, and preventing the introduction of foreign vermin.[62] This view clearly reflects the etiological shift described by Jansen: from a holistic approach to a pathogen-oriented view that centered its attention on the surveillance of vermin and pathogens and tended to disregard the significance of the environment in this matter.[63] This change originated in the field of human medicine, but quickly spread to other sciences such as entomology, botany, and plant pathology.

Phytopathology was primarily focused on studying and understanding the diseases of plants cultivatable in Europe and the United States (for example, wheat, potatoes, canola or rapeseed, wine grapes, apples, pears, cucurbits, and sugar beets). The threat of spreading plant pathogens through global trade, as well as the ever-growing demand for agricultural crops and foodstuffs from colonial tropical and subtropical regions (for example, cotton, rubber, palm oil, coffee, and sugar cane), led to the internationalization and expansion of the field of phytopathology. The botanical garden and research station in Buitenzorg provide an outstanding example of this correlation. This also applies to Wilhelmine Germany, where phytopathology, tropical botany, and topical agriculture became part of the curricula of colonial schools and university faculties.[64] Otto Warburg, a leading collaborator in the KWK, was professor of tropical botany and plant pathology at the Seminar für Orientalische Sprachen (seminar for oriental languages) at the University of Berlin. Other botanists with colonial experience taught at the universities of Breslau and Göttingen, at the Kolonialinstitut in Hamburg (colonial institute), or the Deutsche Kolonialschule Witzenhausen (colonial school) near Kassel. This shift is not surprising if we recall the main purpose of the German colonial enterprise, which was to achieve supply autarky and thus economic independence from other nations. Following this notion, the development of the colonies was thought to have to rely on scientific methods and principles, as they were the only means by which Wilhelmine Germany, a colonial latecomer, could overcome its colonial backwardness.[65] Thus, the study of plant pathogens and the development of means to combat them were a paramount concern for colonial authorities, the KWK, and plant pathologists, among them Walter Busse.

Plant Pathology and Plant Hygiene in the Colonies

When dealing with nonhuman actors, several difficulties arise from the available sources and literature on the subject. Nearly all the material accessible is entirely related to their negative effect on host plants, or advocates actions to counteract or control the agents in question. Consequently, we know a great deal about the mating ability, longevity, and fecundity of, for example, *Earias insulana*,[66] but we know next to nothing about the significance and role of the cotton spotted bollworm in creating and maintaining a functioning, complex ecological network. Historians' understanding would benefit greatly from a shift in perspective. A broader approach that encompasses not only the negative effects of certain species on cash crops but also understands them as elements of a complex network of interrelations and interactions would contribute to de-centering a still prevailing human anthropocentrism.

The common names given to certain insects underline the narrow perception that dominates the colonial sources. Insects that were considered pests were given common names that related to the cash crop they harm. Examples are the coffee berry borer (*Hypothenemus hampei*), the cacao mirid (*Sahlbergella singularis*), or the maize stalk borer (*Busseola fusca*). In spite of the problems of the colonial archive, the available sources are still valuable because they help to analyze the genealogies of human/nonhuman relationships, and they provide important insights into the habitat, food sources, developmental stages, and ontogeny of both human and nonhuman actors.

Busse, not acquainted with the geography, vegetation, or weather patterns of the places he visited in West Africa, relied on information published by other explorers who had visited these areas, as well as information provided by locals and other resident personnel, including government officials, agronomists, and planters during his expedition. Back in Berlin, he connected with a network of institutions and experts from different fields. They studied the specimens he had collected, determined their respective place in the newly established order of things, and finally classified them as plant pathogens—organisms that (only) cause disease. In doing so, Busse and other scientists partially obscured their potential worth and role as, for example, regulatory elements in a tropical or rainforest environment. A broad range of moths, beetles, fungi, and other nonhuman actors were defined as negative elements, harmful to a profit-oriented, monoculture, cash crop plantation environment.

When Busse visited the colony in 1904, he found that the cotton varieties introduced by the Tuskegee experts and grown in monoculture plantations had been dramatically affected by moths, beetles, and fungi. The cotton cultivated by the Ewe in mixed culture, by contrast, remained widely unaffected. As noted above, cotton had been grown, spun, and woven for generations by the Ewe for their own usage. The bark beetles (*Tomicus* group), grubs of the cotton spotted bollworm (*Earias insulana*), red cotton bugs (cotton stainer, *Pyrrhocoridae*), cotton seed bugs (*Oxycarenus hyalinipennis*), and leafhoppers (*Cicadellidae*) that Busse listed in his surveys, were all thriving on the introduced cotton. The maize introduced by the Tuskegee expedition was populated by moths' grubs (*Earias insulana*), while the locally grown corn remained unaffected as well.[67]

As far as cocoa was concerned, Busse mentioned the cacao mirid (*Sahlbergella singularis* Hagl.) as the most hazardous menace to the Cameroon plantations.[68] This insect, native to Central and Western Africa, can naturally be found on the host plants cola (*Cola nitida*), kapok (*Ceiba pentandra*), and red-flowered silk cotton tree (*Bombax buonopozense*).[69] However, it also feeds on pods and shoots of cocoa. Cacao mirids have sucking mouth parts and "feed by injecting saliva that contains enzymes capable of digesting the cell walls and cell contents. This saliva is also phytotoxic and necrosis develops in the tissues surrounding the feeding site,"[70] which in turn leads to infested cocoa trees quickly becoming nonproductive. Furthermore, its presence in cocoa plantations has led to significant production shortfalls of approximately 25 to 40 percent in the Ivory Coast, Ghana, and Cameroon.[71] In his report, Busse stated that he was unable to find out why the mirid had appeared on certain plantations and not on others, and that he had found no solution to the problem besides employing various chemical substances (pesticides) and burning all of the infected plant material.[72]

In contrast to cotton or cocoa, rubber can be harvested from a number of different plants: trees, liana, shrubs, and herbs. Rubber-yielding plants are distributed globally. They are found in the equatorial tropical rainforests, in the arid and semiarid areas in the southwestern United States, North Central Mexico, and regions with similar climates around the world. As mentioned above, in central and western Africa rubber is abundant in *Landolphia* liana and *Kickxia elastica* trees. Other species were introduced by inter- or trans-imperial networks in order to be grown in plantations, such as *Hevea brasiliensis, Manihot glaziovii,* and *Castilloa elastica* from Central and South America, and *Ficus elastica* from Southeast Asia. Of these varieties, *Castilloa elastica* was most affected by the longhorn beetle *Inesida leprosa,* native to central and western Africa. As in the case of the cacao mirid or the corn stalk borer, this beetle is commonly known as castilloa borer. In Cameroon and other parts of West Africa, the *Castilloa* plantations were devastated by the larvae of this beetle.[73] As these insects habitually select sickly or overshadowed trees in preference to healthy ones, their presence can be interpreted as a sign of scarcity or fault in cultivation.[74] Other varieties, by contrast, like *Hevea, Manihot,* or *Ficus,* were insignificantly affected by this insect or any other sort of moth, beetle, or fungi.

The Impact of Plant Pathogens and Phytopathology

The longhorn beetle had a direct effect on the rubber-yielding varieties cultivated in Cameroon. In the same way, the cacao mirid and the array of bugs affecting cotton growth in Togo significantly impacted the course of German tropical agriculture. This is reflected in the articles published in *Der Tropenpflanzer* on the development of German plantations in Africa. The articles reported that in some plantations the cultivation of certain cash crop varieties was reduced. In other cases, the cash crops were replaced by others, and in some cases the cultivation was abandoned completely, like in the case of *Castilloa elastica*.[75] The cultivation of rubber, cotton, and cocoa in Cameroon and Togo collided with the behavior of local moths, beetles, and fungi. These tiny nonhuman actors became entangled in the network of tropical botany and economic agriculture pursued by German colonial policy. The prosperity of the German plantation economy did not only depend on the capacity and knowledge of the planters, the quality of the soil, and weather and rain distribution patterns but also was determined by the activities of numerous small creatures.

Busse's expedition was important because it was the first one to locate, study, identify, catalog, and devise means to counteract the pathogens affecting the plantations. Busse considered the results of his expedition and work "merely as an introduction, as the beginning of the work to come."[76] Because the measures he proposed did not meet expectations, the KWK launched another expedition to Cameroon in 1907. It was led by the botanist Friedrich Carl von Faber (1880–1954),[77] also assigned to study and remedy the ailments in the rubber and cocoa plantations. Like Busse, he was unable to effectively control or devise effective means to counteract the negative effects of nonhuman actors upon the afflicted cash crops. His was also the last phytopathological expedition.

It is not easy to come to a clear-cut conclusion about how to interpret Busse's expeditions and reports in the light of current research. On the one hand, the German expeditions in Togo and Cameroon fit perfectly with the classical narrative of European colonization, discovery, and economic exploitation of the non-European world. On the other hand, however, Busse's reports can be read as the story of a man whose civilizing mission was ruined by insects and bacteria. The scientists anxiously focused their research on these so-called pathogens. A new scientific discipline, phytopathology, was created, and hundreds and thousands of pathogens were identified, named, and classified. The fight against vermin with pesticides and other chemical weapons began. The making of phytopathology was, up to a certain extent, a reaction to a situation in which European scientists realized that they were no longer the superior actors in the transformation of colonial nature. The smooth success story of scientific progress can be turned into a tale of slightly nervous scientists who tried to cope with the unpredictabilities of nonhuman agency.

Notes

1. Walter Busse, "Ueber die Aufgaben des Pflanzenschutzes in den Kolonien," in *Verhandlungen des Deutschen Kolonialkongresses 1905 zu Berlin*, edited by committee (Berlin: Verlag von Dietrich Reimer, 1906), 30–44.

2. "Einheimische Insekten, die vorher auf den Pflanzen des Urwaldes und der Steppe ihr Dasein fristeten, haben sich gierig auf gewisse, ihnen mehr zusagende, eingeführte Nutzpflanzen gestürzt, und sich . . . in beängstigender Weise vermehrt," translated by SEW; cf. Busse, "Ueber die Aufgaben des Pflanzenschutzes," 32.

3. "Das Fortbestehen wichtiger Kulturen geradezu in Frage gestellt wird," translated by SEW; cf. Busse, "Ueber die Aufgaben des Pflanzenschutzes."

4. Walter Busse, "Bericht über die Pflanzenpathologische Expedition nach Kamerun und Togo 1904/1905," *Beihefte zum Tropenpflanzer* 7, no. 4 (Berlin: E. S. Mittler & Sohn, 1906), 163; Busse, "Reisebericht der pflanzenpathologischen Expedition des Kolonial-Wirtschaftlichen Komitees nach Westafrika," *Der Tropenpflanzer* 9, no. 1 (1905): 25–37.

5. On German geographers and the expeditions undertaken ahead of German colonial outreach, see the article by Ulrich van der Heyden, "Die geographische Entdeckung Afrikas und der deutsche Kolonialismus," in *Kolonialismus hierzulande: Eine Spurensuche in Deutschland*, edited by Ulrich van der Heyden and Joachim Zeller (Erfurt, Germany: Sutton, 2007), 99–103; on German plans for developing African territories, see Dirk van Laak, *Imperiale Infrastruktur: Deutsche Planungen für die Erschließung Afrikas* (Paderborn, Germany: Ferdinand Schöningh, 2004), 11–14, 35–37; on German natural scientists and their contributions to German colonial exploration, see Brigitte Hoppe, "Naturwissenschaftliche und zoologische Forschungen in Afrika während der deutschen Kolonialbewegung bis 1914," *Berichte zur Wissenschaftsgeschichte* 13 (1990): 193–206; Bernhard Zepernick, "Zwischen Wirtschaft und Wissenschaft—die deutsche Schutzgebiets-Botanik," *Berichte zur Wissenschaftsgeschichte* 13 (1990): 207–17.

6. Following the abolition of the transatlantic slave trade in 1807 and the subsequent search by European merchants for legitimate commercial relations with West Africa, first oil-palm fruits, later rubber, and subsequently other tropical cash crops became sought-after commercial alternatives. The term "legitimate commerce" was used to designate trade in anything other than slaves, including nonagricultural commodities such as gold and ivory. In practice, it was centered on the commercialization of products obtained from commercial agriculture. Furthermore, it was thought that West Africa could take the place of the Americas as a supplier of tropical products to Europe, with African labor retained and employed locally; Samuel Eleazar Wendt, "Hanseatic Merchants and the Procurement of Palm Oil and Rubber for Wilhelmine Germany's New Industries, 1850–1918," *European Review* 26 (2018): 432.

7. Hartmut Pogge von Strandmann, *Imperialismus vom Grünen Tisch: Deutsche Kolonialpolitik zwischen wirtschaftlicher Ausbeutung und "zivilisatorischen" Bemühungen* (Berlin: Ch. Links, 2009), 48.

8. Richard V. Pierard, "A Case Study in German Economic Imperialism: The Colonial Economic Committee, 1896–1914," *Scandinavian Economic History Review* 16 (1968): 163–64.

9. Of the seventy-six men who served on the committee's board of directors between 1896 and 1914, forty-three were businessmen, fourteen were academicians or professors, twelve were lawyers, judges, and government officials; there was one physician, one explorer, one artist, and four men whose occupations could not be determined. See Pierard, "A Case Study in German Economic Imperialism."

10. In 1909, the KWK published a report about colonial economy, which was based heavily on official statistical material. It sought to inform its readers about the significant increase in trade with the colonies and the importance of tropical raw materials for industrial manufacture, highlighting the economic interdependence between colonial trade and domestic labor conditions. The seven chapters following the introduction are dedicated to different raw materials and arranged in descending order, depicting the significance and value of each commodity for Wilhelmine industry. The first chapter is dedicated to cotton, the second to rubber, and the third to oil palm fruits and kernels. The remaining chapters deal with tropical timber and tanning agents, mineral resources, animal products, colonial edibles, and stimulants such as coffee and cocoa; Kolonial-Wirtschaftliches Komitee, ed., "Unsere Kolonialwirtschaft in ihrer Bedeutung für Industrie und Arbeiterschaft," in *Beihefte zum Tropenpflanzer* 13, no. 3 (Berlin: E. S. Mittler & Sohn, 1909).

11. A detailed account of the developments made in tropical botany and tropical plantations, centered on the developments made in Britain, France, and the Netherlands during the nineteenth and early twentieth centuries, can be found in Daniel R. Headrick, *The Tentacles of Progress: Technology Transfer in the Age of Imperialism, 1850–1940* (New York: Oxford University Press 1981), 209–58.

12. Friedrich Karl Timler and Bernhard Zepernick, "German Colonial Botany," *Berichte der Deutschen Botanischen Gesellschaft* 100 (1987): 147–50.

13. Bradley Naranch, "Between Cosmopolitanism and German Colonialism: Nineteenth-Century Hanseatic Networks in Emerging Tropical Markets," in *Cosmopolitan Networks in Commerce and Society 1660–1914*, edited by Andreas Gestrich and Margrit Schulte Beerbühl (London: German Historical Institute, 2011), 99–132; Reiner Hering, *Konstruierte Nation: Der Alldeutsche Verband 1890 bis 1939* (Hamburg, Germany: Christians, 2003); Hans Jaeger, *Unternehmer in der deutschen Politik* (Bonn, Germany: Röhrscheid, 1967).

14. Buitenzorg's exemplary function can be traced in numerous publications, most prominently in an article written by the tropical botanist and Buitenzorg attendee Otto Warburg in 1899, published in *Der Tropenpflanzer*. Here, the author demanded that a scientific-technical laboratory should be established in the research station of the botanical garden of Victoria, Cameroon, following the example of Buitenzorg. He argued that in Buitenzorg, justice was done to the fact that the field of tropical agricultural science had diversified into a broad variety of specialized fields of study. Consequently, the study of tropical cash crops and their diseases could no longer be satisfactorily tackled by horticulture alone, but instead required different disciplinary approaches, scientific experiments, herbaria, libraries, and periodicals for the pub-

lication and dissemination of findings; in short, a modern scientific infrastructure. See Otto Warburg, "Warum ist die Errichtung eines wissenschaftlich-technischen Laboratoriums in dem botanischen Garten zu Victoria erforderlich?," *Der Tropenpflanzer* 3 (1899): 292; Daniel R. Headrick, "Botany, Chemistry, and Tropical Development," *Journal of World History* 7 (1996): 5; Suzanne Moon, "Technology and Ethical Idealism: A History of Development in the Netherlands East Indies," *Studies in Overseas History* no. 9 (Leiden: CNWS Publications, 2007), 40.

15. Johannes Mildbraed, "Walter Busse—Nachruf," *Berichte der Deutschen Botanischen Gesellschaft* 51 (1933): 61–71.

16. Annerose Menninger, "Die Verbreitung von Schokolade, Kaffee, Tee und Tabak in Europa (16.–19. Jahrhundert): Ein Vergleich," in *Chocolat Tobler: Zur Geschichte der Schokolade und einer Berner Firma*, edited by Yvonne Leimgruber, Patrick Fenz, and Roman Rossfeld (Bern, Switzerland: Historisches Institut der Universität Bern, 2001), 28–37.

17. Francesco Chiapparino, "Von der Trink- zur Eßschokolade: Veränderungen eines Genußmittels zwischen dem 19. und dem beginnenden 20. Jahrhundert," in *Essen und kulturelle Identität: Europäische Perspektiven*, edited by Eva Barlösius, Gerhard Neumann, and Hans J. Teutenberg (Berlin: Akademie Verlag, 1997), 387–400.

18. Annerose Menninger, *Genuss im kulturellen Wandel: Tabak, Kaffee, Tee und Schokolade in Europa* (16.–19. Jahrhundert) (Stuttgart: Franz Steiner Verlag, 2001), 355–72; Margrit Schulte Beerbühl, "Faszination Schokolade: Die Geschichte des Kakaos zwischen Luxus, Massenprodukt und Medizin," *Vierteljahresschrift für Sozial- und Wirtschaftsgeschichte* 95 (2008): 410–29. Roman Rossfeld's article provides an insightful analysis of the relationship between chocolate (whole milk and dark) consumption and gender role ascription: Roman Rossfeld, "Vom Frauengetränk zur militärischen Notration: Der Konsum von Schokolade aus geschlechtergeschichtlicher Perspektive," in *Chocolat Tobler: Zur Geschichte der Schokolade und einer Berner Firma*, edited by Yvonne Leimgruber, Patrick Fenz, and Roman Rossfeld (Bern, Switzerland: Historisches Institut der Universität Bern, 2001), 55–65.

19. Franz Eulenburg, "Konsum einiger Genussmittel," *Weltwirtschaftliches Archiv* 11 (1917): 332.

20. Menninger, *Genuss*, 235.

21. Thaddeus Sunseri, "The Baumwollfrage: Cotton Colonialism in German East Africa," *Central European History* 34 (2001): 35.

22. Sven Beckert, *Empire of Cotton: A Global History* (New York: Knopf, 2014), 355.

23. Sunseri, "The Baumwollfrage," 36.

24. Emilie Carreón Blaine, *El olli en la plástica mexica: El uso del hule en el siglo XVI* (Mexico City: Universidad Nacional Autónoma de México, 2006).

25. For more insights into the uses of rubber and vulcanization equivalents developed by indigenous people in Central and South America, see Jens Soentgen, "Die Bedeutung indigenen Wissens für die Geschichte des Kautschuks," *Technikgeschichte* 80 (2013): 295–324.

26. Julio Calvo Pérez, "Article: Caucho," in *Diccionario Etimológico de Palabras del Perú*, edited by Julio Calvo Pérez (Lima, Peru: Editorial Universitaria Ricardo Palma, 2014), 164–65.

27. Francisco Javier Ullán de la Rosa, "La era del caucho en el amazonas (1870–1920): Modelos de explotación y relaciones sociales de producción," *Anales—Museo de América* 12 (2004): 183–204.

28. Eckhard Kupfer, "Amazonien—vom Kakao über Kautschuk zum hightech," in *Amazonien: Weltregion und Welttheater*, edited by Willi Bolle, Marcel Vejmelka, and Edna Castro, Reihe Lateinamerika-Studien, no. 1 (Berlin: Trafo, 2010), 195–214; Ullán de la Rosa, "La era del caucho," 184.

29. Quentin R. Skrabec, *Rubber: An American Industrial History* (Jefferson, NC: McFarland & Company, 2014), 31–33.

30. Wolfgang Jünger, *Kampf um Kautschuk* (Leipzig, Germany: Goldmann, 1942), 202; Jürgen Ellermeyer, *Gib Gummi! Kautschukindustrie und Harburg* (Bremen, Germany: Edition Temmen, 2006), 15.

31. Cornelius Torp, *Die Herausforderung der Globalisierung. Wirtschaft und Politik in Deutschland 1860–1914*, Kritische Studien zur Geschichtswissenschaft, no. 168 (Göttingen, Germany: Vandenhoeck & Ruprecht, 2005), 147–69.

32. Patrick O'Brien and Geoffrey Pigman, "Free Trade, British Hegemony and the International Economic Order in the Nineteenth Century," *Review of International Studies* 18 (1992): 104; Sidney Pollard, *Typology of Industrialization Processes in the Nineteenth Century* (London: Hardwood Academic Publishers, 1990), 55.

33. Corey Ross, "The Plantation Paradigm: Colonial Agronomy, African Farmers, and the Global Cocoa Boom, 1870s–1940s," *Journal of Global History* 9 (2014): 52.

34. Ross, "The Plantation Paradigm," 51.

35. Schulte Beerbühl, "Faszination," 414.

36. Walter Stollwerck, *Der Kakao und die Schokoladenindustrie: Eine wirtschaftsstatistische Untersuchung* (Halle, Germany: Lippert & Co., 1907), 31.

37. Stollwerck, *Der Kakao und die Schokoladenindustrie*.

38. Stollwerck, *Der Kakao und die Schokoladenindustrie*, 28.

39. Beckert, *Empire*, 355.

40. For further details on the history of the Tuskegee experts and the transnational implications of the Togo venture, see Andrew Zimmerman, *Alabama in Africa: Booker T. Washington, the German Empire, and the Globalization of the New South* (Princeton, NJ: Princeton University Press, 2010).

41. Sven Beckert, "From Tuskegee to Togo: The Problem of Freedom in the Empire of Cotton," *Journal of American History* 92, no. 2 (2005): 511.

42. S. H. Bederman, "Plantation Agriculture in Victoria Division, West Cameroon: An Historical Introduction," *Geography* 51, no. 4 (1966): 354; Dieudonné Mouafo, "La production camerounaise de caoutchouc naturel: Evolution et perspectives de commercialisation," *Canadian Journal of African Studies* 26, no. 1 (1992): 97; Tobias J. Lanz, "The Origins, Development and Legacy of Scientific Forestry in Cameroon," *Environment and History* 6, no. 1 (2000): 104–5.

43. During his expedition, Schlechter published a brief article in the KWK's journal *Der Tropenpflanzer* in 1899, where he gives a broad account of the results of the expedition and indicates that 30,000 Kickxia seeds were brought to Cameroon, out of which approximately 90 percent germinated and were distributed to the plantations and the botanic garden in Victoria. See Rudolf Schlechter, "Die Überführung der

Kickxia von Lagos nach Kamerun," *Der Tropenpflanzer* 3, no. 8 (1899): 357; a full report of his expedition was published in 1900 in the KWK's own publishing house. See Rudolf Schlechter, *Westafrikanische Kautschuk-Expedition 1899/1900* (Berlin: Verlag des Kolonial-Wirtschaftlichen Komitees, 1900).

44. Th. Loesener, "Rudolf Schlechters Leben und Wirken," *Notizblatt des Königlich botanischen Gartens und Museums zu Berlin* 89, no. 9 (1925): 916.

45. George N. Agrios, *Plant Pathology* (Boston: Elsevier Academic Press, 2005), 8–27.

46. Following Foucault, the order of things relates to conditions of truth and discourse, paradigms that provide the epistemological framework and determine what is possible or acceptable to affirm. Michel Foucault, *The Order of Things: An Archaeology of the Human Sciences* (New York: Vintage Books, 1994).

47. P. D. Sharma, *Plant Pathology* (Oxford: Alpha Science International, 2006), 1–14.

48. Hans Braun, "Geschichte der Phytomedizin," in *Handbuch der Pflanzenkrankheiten I, Die nichtparasitären Krankheiten*, edited by Hans Braun and Otto Müller (Berlin: Paul Parey, 1965), 21.

49. Agrios, *Plant Pathology*, 17.

50. Sharma, *Plant Pathology*, 1–15.

51. George M. Reed, "Phytopathology—1867–1942," *Torreya* 43, no. 2 (1942): 156.

52. Herbert Hice Whetzel, *An Outline of the History of Phytopathology* (Philadelphia: W. B. Saunders Company, 1918), 42.

53. Agrios, *Plant Pathology*, 18–20.

54. Reed, "Phytopathology," 162–63.

55. David Bardell, "The Tobacco Mosaic Virus: Reflections on the 100th Anniversary of the Discovery of the First Virus," *Science Teacher* 64, no. 9 (1997): 28.

56. Agrios, *Plant Pathology*, 23–26; on the discovery of prions in plants, see Sohini Chakrabortee et al., "Luminidependens (LD) is an Arabidopsis Protein with Prion Behavior," *Proceedings of the National Academy of Sciences of the United States of America* 113, no. 21 (2016): 6065–70.

57. Sarah Jansen, *Schädlinge: Geschichte eines wissenschaftlichen und politischen Konstrukts, 1840–1920* (Frankfurt: Campus Verlag, 2003).

58. Jansen, *Schädlinge*, 14.

59. Tobias J. Lanz, "The Origins, Development and Legacy of Scientific Forestry in Cameroon," *Environment and History* 6, no. 1 (2000): 100.

60. Jansen, *Schädlinge*, 11–17; Lukas Straumann, *Nützliche Schädlinge: Angewandte Entomologie, chemische Industrie und Landwirtschaftspolitik in der Schweiz, 1874–1952* (Zürich, Switzerland: Chronos, 2005); Jana Sprenger, "'Die Landplage des Raupenfraßes' Wahrnehmung, Schaden und Bekämpfung von Insekten in der Forst- und Agrarwirtschaft des preußischen Brandenburgs (1700–1850)" (PhD diss., University of Göttingen, 2011).

61. "Es macht sich hier [beim Pflanzenschutz, SEW] dieselbe Erscheinung geltend, die auch beim Kulturmenschen und den Haustieren schon seit langer Zeit hervorgetreten ist, und die in hygienischen Maßnahmen aller Art ihren beredten Ausdruck findet," translated by SEW; cf. Busse, "Aufgaben des Pflanzenschutzes," 30.

62. Busse, "Aufgaben des Pflanzenschutzes," 36–37.

63. Jansen, *Schädlinge*, 277.

64. Friedrich Karl Timmler and Bernhard Zepernick, "German Colonial Botany," *Berichte der Deutschen Botanischen Gesellschaft* 100 (1987): 149–50.

65. Charles Theodor Hagberg Wright, "German Methods of Development in Africa," *Journal of the Royal African Society* 1, no. 1 (1901): 23–38, 33–34.

66. M. Kehat and D. Gordon, "Mating Ability, Longevity and Fecundity of the Spiny Bollworm, Earias insulana," *Entomologia Experimentalis et Applicata* 22, no. 3 (1977): 267–73; M. Kehat et al., "Sex Pheromone Traps as a Potential Means of Improving Control Programs for the Spiny Bollworm, Earias Insulana," *Phytoparasitica* 9, no. 3 (1981): 191–96; Ashok J. Tamhankar, "Host Influence on Mating Behaviour and Spermatophore Reception Correlated with Reproductive Output and Longevity of Female Earias Insulana (Lepidoptera: Noctuidae)," *Journal of Insect Behaviour* 8, no. 4 (1995): 499–511.

67. Busse, *Pflanzenpathologische Expedition*, 204–15.

68. Busse, *Pflanzenpathologische Expedition*, 178.

69. Dennis Leston, "Entomology of the Cocoa Farm," *Annual Review of Entomology* 15 (1970): 279.

70. T. W. Tinsley, "The Ecological Approach to Pest and Disease Problems in West Africa," *Journal of the Royal Society of Arts* 112, no. 5093 (1964): 357.

71. Régis Babin, "Cocoa mirid control recommendations," http://www.cirad.fr/en/research-operations/research-results/2009/cocoa-mirid-control-recommendations.

72. Busse, *Pflanzenpathologische Expedition*, 179.

73. "Insects Destructive to Cultivated Plants in West Africa," *Bulletin of Miscellaneous Information (Royal Gardens, Kew)* 1897, no. 125/126 (1897): 175–91; Small Ovoid, "A Project That Failed," *Empire Forestry Review* 25, no. 2 (1946): 240–47.

74. "Insects Destructive to Cultivated Plants in West Africa," 180.

75. Selig Eugen Soskin, "Die Internationale Kautschuk-Ausstellung in London, 14–26 September 1908," in *Beihefte zum Tropenpflanzer* 12, no. 11 (Berlin: E. S. Mittler & Sohn, 1908), 319.

76. "Meine Untersuchungen über die Natur der fraglichen Krankheiten, über die Lebensbedingungen der pflanzlichen und tierischen Schädlinge, und die noch im Gange befindlichen Versuche zu deren Bekämpfung können nur als eine Einleitung, als Beginn der Arbeit angesehen warden," translated by SEW; cf. Busse, "Aufgaben des Pflanzenschutzes," 33.

77. Friedrich Carl von Faber, "Bericht über die pflanzenpathologische Expedition nach Kamerun," *Der Tropenpflanzer* 11, no. 11 (1907): 755–77.

French Mandate Syria and Lebanon

Land, Ecological Interventions, and the "Modern" State

IDIR OUAHES

The case of agricultural experimentation in the early French mandate period demonstrates the shift of ecological interventions by organized state actors from romantic orientalism to coherent state-centered development after the First World War. The interwar era witnessed an increasing focus on state-building logics in agricultural affairs. This case study on French Mandate Syria and Lebanon demonstrates how agrarian doctrines of development informed a policy of science, measurement, and control. The history of agrarian reforms in Syria and Lebanon builds on—as well as critiques—historical interpretations of agrarian development in the French Empire and the British Empire. The case study of agrarian efforts—ranging from species introductions and fertilizers to legislation—examines hitherto unknown interactions between the local population, the French, and neighboring colonies and countries relating to agrarian improvement.

France took over Syria and Lebanon from the British military administration in 1918, after the collapse of Ottoman rule over the Levant. France "received" Syria through the League of Nations Mandate, with a stipulation that France should exercise tutelage of the local peoples to prepare them for independence. France's League of Nations Mandate framework required a rapidly assembled state-building exercise. France brought officials with extensive experience in French colonial North Africa to help develop a modern colony in Syria.[1] France received official control over Syria only in 1922, although French direct rule over the region had begun in 1918 in the Mediterranean coastal regions of Lebanon and Lattakia, and in the summer of 1920 over the rest of Syria "proper," which had sought to form an independent Arab Kingdom under Hashemite leader Faisal Bin Hussein Bin 'Ali Al-Hashimi.

To incoming colonial administrators, Syria and Lebanon could be understood as regions whose people and nature needed "improvement" to fulfill colonial agrarian ideals. Syria and Lebanon were "colonies" of Europe despite the humanitarian language of the League of Nations. The evolution of the modern nation state, rooted within European and colonial experiences, undertook increasingly sophisticated efforts to shape nature and society using science, technology, and measurement.[2] Imperial domains offered more malleable testing grounds for ecological interventions because autocratic powers associated with colonial rule gave experts and state officials greater ability to imagine and implement plans for reformatting nature and society. Joseph Hodge's review of science in the British Empire noted that power was unevenly exercised through scientific knowledge and innovation in British colonies.[3] The increasing interest in the role of state-directed ecological interventions in the British Empire was paralleled in the French Empire. Diana Davis has dealt with the role of the French colonial forest service in shaping the environment in Morocco. She examined how the creation of protected forests resulted in the dispossession of local peasants.[4] Discussing a similar subject, Caroline Ford notes that the case of forestry in French Algeria demonstrated "the darker sides of environmentalism" because of its dispossession of peasants.[5]

Post–First World War developments in Syria and Lebanon built on late nineteenth-century colonial efforts to manage the people and nature of French North Africa. French colonial policy increasingly emphasized the *mise en valeur* (literally meaning "to make worthy") of colonial territories. This concept was replacing the earlier romantic-orientalist rhetoric and representation that had tended to idealize the "pure" and "guileless" exotic "Other" people and environments to be found in the Middle East, Africa, and Asia. It is notable that, in contrast to the presence of orientalist and romantic imaginations in the German and British agricultural services, French savants and experimental researchers in such places as Algeria had initially focused on the aesthetics of a civilizing mission rather than questioning the degree of economic and technical success achieved.[6] The concept of giving economic value to colonial possessions, which paralleled the "constructive imperialism" outlined earlier in Britain by Colonial Secretary Joseph Chamberlain, was one rung up in a "ladder" that tied the romantic-era "civilizing mission" of the nineteenth century to the developmental state of the interwar and early post–Second World War periods.[7]

This chapter situates the history of agriculture within wider historiographies focused on the environment, science, and imperialism.[8] Colonial agricultural efforts represented the pinnacle of what James Scott described as "high modernism," which sought to control people and nature for human benefit. Scott argues that: "Agriculture is, after all, a radical reorganization and simplification of flora to suit man's goals."[9] Scott's emphasis on the state also allows him to see the particular force and direction given to ecological interventions in the modern period, often starkly evident in colonial circumstances.[10] Situations where a state-imposed logic was intended to improve agricultural production had cascading effects as a result of ecological and human reactions. Timothy Mitchell has noted that the Egyptian state's policies encouraging a "free-market" agriculture in fact led the peasants to plant cash crops to survive; a reaction that carried significant aftereffects for the country's economy and ecology.[11]

At the same time, we should be wary of casting the state as an all-seeing and powerful agent that harmed the interests of locals. In many respects, French administrators had little power over rainfall, heat, local farming practices, and fertility. Nor did their efforts always meet with criticism. Local Syrian and Lebanese elites agreed with French agrarian improvement efforts and the introduction of useful plants and animals. This point has a wider relevance for the French Empire in North Africa and the Middle East. Interwar agricultural efforts did not simply conform to Diana Davis's suggestion that a declensionist narrative emphasizing a lost agricultural potential owing to Arab deforestation retained its appeal in the twentieth century. Writing in a specialist journal in 1931, French-Algerian agronomist Jean Blottière presented a statistics-informed outline of the growth of cereal agriculture and even noted that Algeria had been a net exporter of grains prior to the French conquest, an endeavour that Blottière admitted had harmed indigenous cereal production.[12] The same edition of *L'Agriculture des Pays Chauds* carried an account of cotton growing in the country, which noted that cotton had been introduced by the Arab conquerors, with the root of the word "cotton" itself coming from the Arabic *qtun*.[13] The purpose of this chapter is to illuminate the historical development and contradictions of French agrarian efforts in mandate Syria and Lebanon.

The Levant Context

France was something of a latecomer in the Levant. The Levant had experienced Middle Eastern state interventions in agricultural and water management from antiquity onward; a phenomenon (in)famously described as "hydraulic despotism."[14] The Middle East was also noted to have experienced an "Islamic Green Revolution" in the early days of unified Muslim rule. Edmund Burke III suggests that the reversal of this agricultural growth was the result of a combination of human and environmental factors, including plant diseases, in the medieval period.[15] Such state interventions continued under the Ottomans, though the imperial nature of their rule, its Islamic foundations, and confusing layers of taxation, did not particularly increase state interference in ecological situations. As late as 1886, Ottoman land was divided into five categories, with different categories of taxation forming an even more complex postfeudal system. This complex system limited the Ottoman efforts at agrarian modernization and complicated later French efforts.

First was *mülk* land, which was subject to proprietary law according to the civil code. Second were the *waqfs*, which were mortmain (perpetuities with absolute or relative inalienability). Third were the *mirie* state lands. Fourth were the *metruke* lands, which did not have private claims and were instead reserved for public and communal use (such as highways and communal pastures). Finally, there was *mewat*, undeveloped lands. French authorities were most interested in *mirie* lands since they were the most cultivable and accessible to the state. This land was rented on a usufruct (shared) basis. If this type of land had gone uncultivated for three years, however, the rentiers would lose their rights. The lands were rented for either money or a tithe of food. French Mandate agricultural advisor Edouard Achar also noted the existence of great landlords with latifundia that were hampering development and even suggested a progressive weakening of the hold of these landowners and the parceling out of the land to productive farmers.[16]

Despite this labyrinthine organization, there had been some Ottoman efforts at modernization. In the 1880s, Istanbul's Ministry of Trade and Public Works employed a director for agriculture who sent inspectors out to advise farmers on improving their techniques, and Istanbul sent young Ottomans to Europe to study agricultural science. A more professional agricultural policy was outlined during the 1890s, with agricultural schools set up at Muslimiyah near Aleppo, Salonica, and Bursa, as well as experimental farms.[17] An agricultural bank was organized in Istanbul in 1898. This bank had branches in Damascus that had continued to operate during King Faisal's independent Arab government. Branches were opened in Beirut during the First World War, though the *mutassarifate* of Lebanon was not interfered with due to its special administrative autonomy. As they retreated during the First World War, the Ottomans took most of the banks' accounts with them. The Faisalians reopened the agricultural banks and undertook painstaking work to recover the names of the banks' investors by using local administration archives. The bank was thus able to again lend money to farmers for machines, tools, and instruments. Upon reviewing this established system, the agronomic advisor Lieutenant Florimond accepted its core utility, though he suggested that there was a need to unite the funds of the various banks into one central bank for the mandate territory. This would allow the poorer regions to draw on greater loans than their limited local deposits could afford.

Despite these plans for reform, the agricultural situation on the ground was deemed to be dire. The Faisalian government in Damascus estimated in 1918 that the retreating Turkish army had seized 798,269 Turkish lira of deposits from the agricultural banks.[18] By the beginning of the mandate, credit for farmers was being primarily loaned by individual bankers via buyback contracts. The high interest rate on these loans meant that farmers could not pay their debts, leading to foreclosures that allowed banks to consolidate ever larger bank-owned latifundia.[19] In contrast, research on Ottoman agriculture emphasizes the relatively low level of concentrated land holdings in earlier periods.[20] A 1931 review of land laws in Syria by an agronomist attached to the State of Syria's government outlined the French view that the "defectuous" Ottoman land holding system had harmed agriculture.[21] An undated report from the period of France's arrival in Syria outlined agricultural conditions in Syria in the preceding years. It noted that agriculture was the keystone of Syria's prosperity. It estimated that up to 95 percent of the country's locally produced exports were agronomic and accounted for up to 50 percent of the tax income for local governments. The report explained that to "understand Syria, to recognize her proper worth, it is necessary to go beyond the coastal mountain chain . . . one must have visited the great plateaus of fertile land . . . on which millions of hectares remain unexploited."[22] Even at such an early stage, the state-focused, developmental attitude toward agricultural intervention was being outlined.

The Ottoman precedent was changed by an increasingly intrusive mandate state system in the early 1920s. For instance, the State of Syria, founded in 1924, arrogated several types of Ottoman land titles. These included the *mudawara* lands previously held by the Ottoman Sultan, the *mussakafates* (public buildings), the *moutafaouda* (land repossessed by the state), and *mahloule* (goods without inheritance claims), as well as all other holdings through eminent domain and *domaine utile* (rights secured through productive use of the land). Of the mudawara lands, there were 500,000 hectares (approximately 1,235,000 acres) in Aleppo of which four-fifths were being cultivated. In Damascus, there were 1250,000 hectares, of which only 130,000 were cultivated. This came to a total of 1,750,000 hectares, of which 530,000 were developed. Two types of Ottoman taxes were imposed on privately owned land. The first was the land tax, the *werko* based on the worth of the land. The other, a variable one, was the land tithe, *ashour*, which was applied to cultivations. The werko tax was charged at 4 percent of the land's value, though as the land became increasingly dotted with fruit trees, this rose to 12 percent. By the time of the French Mandate, the taxes charged were generally low with respect to the revenue from the lands. This was due to land worth estimates still being based on those of the 1870s, with a significant undervaluation of their worth.

To combat this, French mandate authorities organized a service of land registration, the *Service du Cadastre*.[23] In the State of Syria, the lands were also subjected to inspections (*teftiches*) that were undertaken by an inspector (*moufatache*) and a tax collector, who covered 40,000- to 60,000-hectare regions. This structure was overseen by a centralized land service with two agencies in Aleppo and Damascus. In each agency, there was an agent of the domain, an inspector for the territory, an accountant, a secretary-archivist, a typist, a designer-surveyor, a tax-collector, and an orderly. The agencies were overseen by an audit service, made up of a comptroller, an accountant, a translator-secretary, a typist, and an orderly. The state land generated 230,000 Syrian pounds of tax income in the early 1920s.[24] A 1926 report by authorities in Paris praised the cohesion of the High Commissariat's agronomic advisor, the agricultural services, and the intelligence services in monitoring ecological developments.[25]

Ecological Limits to Mandate State Intervention

Although the Ottoman land system was reformed by the new mandate authorities, the full force and directedness of mandate state capabilities still encountered the limits imposed by ecological actors. Instead of relying on romantic-orientalist discourses on the ineffectiveness of Ottoman agriculture, French efforts in Syria were guided by an attempt to exercise sovereignty over ecological actors. These efforts often encountered mixed results. This was signaled early on when changes were forced upon mandate authorities in Cilicia. By 1918–19, only one-fiftieth of the cultivable lands were being developed. The following year, this had more than doubled. However, climatic conditions, including heavy rains from February to May 1918 followed by a dry summer, caused damage in the 1918–19 year. The drought continued right up to December 1918. The harvest was thus very poor, which, in combination with the arrival of thousands of Christian refugees in Cilicia, caused problems for the region's food supply. French authorities organized monthly meetings of farmers, millers, and exporters to coordinate food supply. However, they miscalculated the scale of the problem, leading to a requisition of barley stocks that was not well received by local farmers.[26]

Weather, pests, and other environmental factors proved especially challenging for mandate authorities in the 1920s. Rains were poor in 1923–24. A particularly cold winter had led to subzero temperatures and snowfall, even on the coast. A freak event—a tornado on June 15—required 10,000 Syrian pounds in compensation to those affected.[27] Cold and drought diminished agricultural output to some extent. The agricultural bank provided 850 quintals (hundredweights) of oat and wheat seeds and 500 quintals of Cypriot winter barley because of the domestic shortfall in production.

These weather conditions had the beneficial effect of killing the cereal larvae (*Dudet al-Zarḥ*). But not all pests suffered. In the spring of 1923, arsenic was spread in the region of Suedieh to fight against voles that "ravaged" fields.[28] In 1924, a fight against another kind of "parasite" was undertaken, targeting a larva called *Dudet Al-Maher*, which affected barley and wheat. The parasite was a microlepidoptera (micromoth) and a technical study was undertaken to determine its habitat and evolution cycle.[29] By 1926, there was a concerted effort at fighting cotton parasites in the region.[30] The northern frontier was even momentarily closed in 1925 to avoid the spread of bovine pests.[31] Reports of poor harvests continued until the late 1920s; one 1928 report noted continuing dry conditions in Syria. So too were grasshopper invasions reported in that year.[32]

This latter ecological actor was a persistent thorn in the side of French agricultural plans, as was the case in other colonial domains.[33] Early in the mandate period, cultivation in the Jabal 'Amil of South Lebanon was noted for having been invaded by grasshoppers, though the impact was less severe than in other parts of the country.[34] The High Commissariat's agricultural service noted that invasions of grasshoppers represented the principal danger to different agricultures. A coordinated fight against parasites was undertaken, with technocrats being made responsible for the destruction of grasshopper nests. These efforts were controlled by specialists of the economic service and even members of the army intelligence corps. Turco-Syrian commissions on the grasshopper threat were organized with Turkey.[35]

Alongside securing the mandate state's raison d'être and modus operandi, ecological interventions were the subject of otherwise rare tutor-tutee consensus. Lebanese newspapers *Al Qa'baa* and *Alef Ba* published a report by the director of agriculture Yousef Atallah regarding locust culling, ensuring that the reading public were informed of governmental efforts at protecting crops.[36] Such reports gave details of efforts in the fight against locusts. In Deir Ez-Zor in mid-April 1930, a report published in *Al-Qabas* described "the fighting . . . against the flying Nejdian and the Moroccan locusts." Near Hama, 34,492 kilos of the flying locusts were killed.[37] The nationalist Beirut newspaper *Al-Sha'ab* wrote an article critical of the governmental response to locusts, which it claimed "fill the plains and the mountains. . . . Rush about like waves." The newspaper reported that a local committee to destroy the locusts was formed to pressure the government for financial aid to fight the insects, an effort that nationalist and future Syrian premier Fakhri Al-Baroudi encouraged. The committee's proposal to incentivize the killing and collection of locusts by paying and fining individuals for participating in or ignoring the culls was adopted by mandate authorities.[38] Pressure on French authorities was increased when the nationalist Baroudi was subsequently sent an official government of Palestine report on Britain's successful locust repression by the British Consul in Damascus.[39]

Competition between the mandate states was short-lived. The consensus on fighting locusts meant that in 1926, French and British mandate authorities reached an agreement with Turkey to organize responses to and prevent locust swarms. The agreement founded an International Bureau to fight locusts, based at Damascus. The Locust Bureau's activities would be guided by a committee formed of the delegates from the states involved. Its primary purpose was the communication of intelligence on the "positioning, extensions and density of oviposition sites (*champs de ponte*), their stage of development, the direction of locust swarms, methods of control and for fighting." The bureau would coordinate this exchange, which would be based on data put together by dedicated special divisions within the respective states' agricultural departments.[40] However, these state-led efforts at controlling ecological movements that threatened crops were not self-evidently successful. For instance, it had taken a month for the French authorities in Aleppo to gather information on locusts. The British Consul noted that the High Commissioner's delegate in that region assumed "a somewhat passive attitude in the matter as he holds that successful methods for dealing with the question are in use in Northern Africa and accordingly it is not considered of any great utility to keep careful statistics."[41]

A French report underlined the mobilization of the mandate state's resources to fight ecological actors. It explained that: "the locust invasion of 1930 was of a particularly dangerous character. . . . Since 1865, several locust invasions have been recorded in Syria [in 1878, 1890, 1902, and 1915]. . . . The locusts that invaded Syria in 1930 belong to two different species. . . . The Nejdi Grasshopper . . . the Moroccan Grasshopper . . . [both are] acclimatized to Syria." The report noted that the Moroccan locusts' eggs could survive throughout the summer, meaning any efforts against them had to cross over into the next season. The report described the "propagation" of the locusts that "continued to advance west . . . crossing the Syro-Lebanese and Syro-Alawite borders . . . to ensure the rapid execution of anti-locust efforts . . . each region was divided into sectors for agricultural defense. Technical officers (graduates of the agricultural schools) . . . have been mobilized . . . to enable technical personnel in each sector to relentlessly pursue the destruction of locusts." Mobile groups of gendarmerie with flamethrower teams were dispatched. Despite these concerted efforts, the locusts were estimated to have destroyed 1.25 to 1.5 percent of Syria's crops.[42]

Alongside pest control, an important introduction of modern state-directed ecological intervention was the increasing use of chemical fertilizers. Chemical fertilizers were little used before the First World War, and the French authorities sought to increase their use under the logic of increasing agricultural productivity. Authorities noted that chemical fertilizers were being used for particular crops, such as *bamieh* (okra) and cotton, although for cereal crops only natural manure was used. One report complained that the "lack of chemical fertilisers prevents the peasants from exploiting to the full the fertility of the soil."[43] Alongside the logic of increased productivity, the use of fertilizers would have an impact on encouraging metropolitan commerce. The *Société Commerciale des Potasses d'Alsace* and other merchants were noted by administrators to be keen to use their potassium fertilizer products in Syria. However, because of their high cost and the exploitation of the soil by sharecropping, these products were judged unlikely to be widely used. Instead, authorities acknowledged that in the short term, the use of "green" fertilizers such as beans, chickpeas, and lentils could be favored. These fertilizers were nothing new to Syrians, who were perfectly aware of the positive impact of intercropping.[44] Certain French actions, such as High Commissariat order number 1449, which banned the export of manure from the Alawite state, encouraged the use of green fertilizers. This was done because the Alawite farmers needed manure for the tobacco and tombac plantations.[45]

Other reports demonstrated less pragmatism from mandate authorities on the issue of fertilizer use. In the *Sanjak* (district) of the *Jabal Barakat* in Hatay province, agriculture was focused on the two plains of Islahiye and Osmaniyah, in the shadow of the Nur mountains. A report expressed surprise that despite their fertility, the two plains were largely fallow. The report pointed out that this could be blamed on a "perfectly uninterested" Ottoman government and the effects of the First World War. However, it noted that many similarly fertile plains in Turkey had seen productivity growth in the first two decades of the twentieth century. Instead, the blame was put on the *longue durée* invasions and population movements of this strategic crossroads between Syria and Anatolia. This had led the local populations to take refuge in the mountains and give up agriculture and was made worse by continuing lawlessness, such as brigandage, which made farming unstable. The report heaped self-congratulatory praise on French authorities for having brought protection and claimed that the locals were slowly being reassured by French security. Fascinatingly, a certain ethnic order was also outlined, which seemed to be attached to abilities to absorb European agricultural methods. The Rumeliotes were deemed good cultivators but materially poor. The Kurds were deemed poor and without a sense for cultivation. The *Çerkes* and Armenians were deemed intelligent and able to make use of modern machinery. The report added that fertilizers were entirely unknown in the region and manure was rarely used and claimed that the farmers were irrationally suspicious of new methods.[46]

To the south, in Lebanon's Jabal 'Amil, agriculture was similarly judged to suffer from the mountainous region and a lack of communications and capital. The peasants had, over centuries, put up foundations that held up the land. However, this had the impact of encouraging deforestation as a result of winter mudslides. The report noted that French machinery was beginning to make an appearance, although it added that peasants "ignored" the utility of the natural and chemical fertilizers. The *seguias* irrigation canals put up by locals to divert water from the local rivers tended to cause water loss. The report also criticized a postfeudal system that meant land was generally owned by great landlords named *beys* or *achaiers* and was worked by the sharecroppers. The landowner provided the funds and animals while the laborer provided seeds, material, and labor. After cultivation, the produce was split, with between one-third and one-half of the cereals going to the landowner.[47]

In their efforts at intervening in the environment, French authorities found local governing elites to be supportive. A report from Greater Lebanon noted that a local administrative commission approved of French methods used to fight diseases of citrus and olive trees and asked for them to be made more widespread. To do this, the Lebanese government sent out a circular notice with methods for fighting diseases.[48] This was in response to a petition sent to the government by local representatives who portrayed the dire impact of citrus diseases. The letter explained that "if this disease continues to spread . . . there is no doubt that that will be the coup de grace for the inhabitants [of Saida's] fortunes . . . one of the landowners of Saida said: 'If the government is not interested in our issue . . . we will tell them to take over our lands and give us free access to the sea; the world is vast.'"[49]

The administrative council of the Lebanese Assembly met to discuss the matter. Fourteen members, presided over by Daoud Bey Ammoun, considered the issue presented before their session by Rachid Bey Jumblatt, Yussef Bey Jouhari, and Nasri Bey Azouri. Jouhari noted that some 1,500 oranges were being lost every day in each plantation. Rachid Jumblatt observed that some 200,000 Syrian pounds were being raised in taxes on the gardens of Sidon. The French governor of Lebanon's representative Francis Petit, justified the efforts being made by the agricultural department of the Lebanese government. Petit added that, in contrast to the department, several local landowners were struck by "inertia" and passivity. The assembly then heard from the Lebanese director of agriculture, a Mr. Younes, who outlined the efforts made to fight diseases. He also noted the unanimous passivity of various notables in the Sidon region, leading the authorities to act alone, sending technicians in cars with the necessary equipment. Regarding the major disease afflicting the olive trees, Younes highlighted that the disease was more difficult to fight since it was caused by an insect that lived between the bark and the branches. It was thus impossible to fight through external methods, and the only remedy was to cut the trees regularly.[50] This case demonstrates how some local elites worked in tandem with French authorities in their efforts to control ecological outcomes. The logic of technocratic state ecological interventions met with less local political resistance than the traditional forms of political arrogation concerned with social, economic, and political matters.

Mandate State-Sponsored Plant Transfers

Mandate state ecological intervention also occurred in transfers of plants. Though this was in the context of longer-term mutual and local ecological exchanges in this crossroads region, a particular organization and direction was given to ecological transfers by the mandate state.[51] This was a result of the colonial state justifying its modus operandi as it sought to develop the country (*mise en valeur*). The long-term dangers of heavy-handed interventions on ecological balance were rarely considered. As James Scott explains "the very strength of scientific agricultural experimentation—its simplifying assumptions and its ability to isolate the impact of a single variable on total production—is incapable of dealing adequately with certain forms of complexity. It tends to ignore, or discount, agricultural practices that are not assimilable to its techniques."[52]

In the local public sphere, there were different opinions about the introduction of species to increase agricultural profits. In 1924, the Beirut newspaper *Al-Watan* wrote that emigration away from the Levant could be reduced by encouraging agriculture.[53] Another newspaper, *Al-Barq*, published an article encouraging the possibility of growing cotton in the country.[54] French government officials reported that local peasants would be interested in accessing the increasing number of seeds available in an agricultural bank.[55] Other public sphere commentaries, however, were more critical. The newspaper *Lisān al-Hāl* wrote that the experimental farms were expensive, with most of their funds going toward technical personnel. The newspaper acknowledged that such highly qualified personnel were necessary for experimentation in cotton culture because this was a new phenomenon in Syria. However, it questioned why they were required for mulberry- or olive-growing cultures since local agriculturalists already had long experience with these. It suggested reducing technical advisors for these two plants and instead using the money to give more seeds to local farmers.[56]

With a dual logic of securing the foundations of domestic and colonial rule through the promotion of French commerce and mandate legitimacy, authorities used their organizational capacity to direct ecological transfers. In 1923, it was reported from Damascus that experimental seed banks had been organized. They included cuttings of a range of seeds.[57] A report noted that sorghum had had excellent results, with the seeds proving to be resistant to poor climatic conditions.[58] Antioch's plant nursery had sold 95,000 cubic feet of mulberries by the early 1920s.[59] Another report noted that although a poor cereal harvest in the Alawite state had resulted from dry weather in the spring, the creation of nurseries in January 1924 had produced promising results in the Tartus region, which had produced 100,000 mulberry trees. The report also noted some successes in the acclimatization of new seeds and an Iranian specialist was hired by the administration to experiment with tobacco cultures. Authorities were adept at combining this with promotion of French capital goods. Demonstration stations were organized to showcase French products such as threshers, straw choppers, and engines.[60]

Cotton was principally cultivated in Idlib and the Jebleh plain, which produced around 2,000 tons. Authorities expected it to develop in Syria since it would fit with France's "efforts . . . to produce in countries submitted to her influence, the primary materials necessary for her factories." Officials expressed their hope that introducing cotton to the Euphrates plains, Alexandretta, Lattakia, and Tripoli could lead to production of 35,000 tons. To encourage this, the High Commissariat organized an experimental station in the 'Akkar plain. In Damascus, a trial of Egyptian cotton was undertaken in 1921.[61] Transplanted cotton seeds were described as having left the experimental stage in 1924. More farmers were seeking to know about experimental cotton seeds, leading to 23,143 hectares being cultivated in 1924, primarily in the Aleppo region. This led to a doubling of the previous year's Aleppo harvest. Cotton had been nonexistent in Damascus in 1924 and very rare in other states. In Aleppo, it was the nonirrigated cotton, named *baladī*, which was most popular with cultivators. According to French authorities, this preference persisted despite the great deal of work that this strand needed from farmers. However, the French sought to encourage the irrigated Egyptian cotton and organized irrigation projects toward this end. The High Commissariat based this on studies of Egyptian cotton's productivity.[62]

Trials of Egyptian and U.S. cotton were regarded as successful and as "perfectly acclimatized to the soil and climate of the Alawite State." Various strains of cotton were grown, including the Egyptian *sakellarides* and *zagora* cotton, as well as *yerli* from Cilicia and the local Levantine *baladī*. The president of the Mulhouse Chamber of Commerce Mr. Dolfus sent a U.S. strain, the Texas Good Middling, in May 1923. This variety was planted in Jebleh by the Aly Dib family. Of 3,000 dunums of land, half of the cotton cultivated was *baladī*, 1000 dunums were Egyptian, and 500 were strains from Cilicia or the United States. The bolls of the U.S. cotton were three to four times bigger than Egyptian cotton and 10 times bigger than Syrian *baladī* cotton, and also opened much faster—within thirty days. However, the Egyptian strain remained the most productive and had the finest and longest fibers.[63] The scope of French imperial bureaucracy was thus particularly influential in encouraging ecological transfers of regional, even global, reach.

The most evident case of this was the employment of French diplomats in Brazil to secure the transfer of a variety of Brazilian seeds for trials in the Levant. In March 1926, the French ambassador to Brazil, A. R. Conty, wrote to the French consul in Bahia, Léon Hippeau. He explained that the agricultural service in the State of Syria under Edouard Achard was interested in paying for samples of kidney cotton (*Gossypium brasiliense*), which was reportedly being cultivated in the states of Alagoas and Sergipe. Kidney cotton was believed to provide harvests within only three months of planting. Conty asked for 10 kilograms of seeds, along with a note on their cultivation and economic worth.[64] The consul passed on the request to a director of the

State Bank of Sergipe.[65] The director then sent a request on the consul's behalf to the Brazilian authorities, seeking 10 kilograms of kidney cotton. In reply, however, the authorities denied the diplomats' claims about the plants' abilities. A member of the agricultural department, Heitor Andre Tavares, explained that the subspecies they had requested did not have the qualities of fast turnaround that they had expected. In fact, kidney cotton was usually harvested after seven to eight months of cultivation. He instead suggested an alternative, "Day's Pedigree," which had a cultivation period of four to five months.[66] After further enquiries, the consul's agent in Maceio, M. C. Girard, informed Leon Hippeau that kidney cotton was not in fact available in the northern states of Alagoas and Sergipe and it only grew in the southern states, mainly in Sao Paolo.[67] Considering this, the consul settled for Day's Pedigree and 10 kilograms were dispatched to Damascus, gaining Achard's gratitude.[68]

Alongside cotton, trials of foreign tobacco were undertaken in the Alawite state and Lebanon. The local tobacco, nicknamed *Abu Riḥa,* was sold to England and the United States. French authorities reported that the regulatory barrier imposed by the Ottoman *Régie des Tabacs* had hurt this crop, which would now prosper at levels not seen since its nineteenth-century heyday, when Syria was the main provider of Egypt's tobacco.[69] Experimentation with tobacco strains was augmented as administrators sought to implement this policy of *mise en valeur* of their territory. In 1924, the production of 300,000 kilograms of local tombac was judged to be a poor performance in contrast to competing Iranian tombac. Use of the local *baladī* tombac had become inefficient, with over three-quarters of the tobacco grains becoming spoiled. Seeds from Shiraz in Iran were imported and distributed to counteract this. Further trials with the Iranian tombac were undertaken in Lattakia, Bouka, and Jebleh.

Tutor-tutee collaboration was again evident in ecological transfers. In the Alawite state, an agricultural bank provided farmers with selected foreign and indigenous seeds. The agricultural chambers were also reported to have supported the administration's efforts at trials.[70] A petition by Alawite notables praised French authorities for their efforts and called for the opening of agricultural and scientific schools, as well as banks in each *qaḍā'* (district) to encourage agriculture.[71] The Alawite region also witnessed organized olive tree transplantation. Several thousand young wild olive trees were planted in the early 1920s. A 500 Syrian pound subvention was paid for the organization of forty plant nurseries, which allowed the distribution of seeds to farmers who were reported to be increasingly interested in shrub culture and had created several private nurseries. A model magnanery was based in Safita and later moved to Banias to demonstrate sericulture techniques.[72]

French efforts at acclimatizing plants to the Syrian situation also focused on forestry in desertified areas. Diana Davis has noted, in her study of French approaches to "desertification" in colonial Algeria, how romantic-orientalist tropes of an Ancient Roman breadbasket shaped administrators' discourses, which laid the blame for North Africa's agricultural decline on short-sighted nomadic pastoralism.[73] In response, the French encouraged planting fast-growing nonnative trees, such as eucalyptus, though these soon dotted the coastline and may have contributed to a decline in biodiversity in the long term. Eucalyptus acclimatization in French North Africa had already been brought to the Levant prior to the mandate. A Mr. Morel ran a "Villa Eucalypta" in Beirut and had begun planting the tree in 1893. Justifying the spread of the trees in the Levant, Morel explained in particularly romantic terms that this tree was strongly suited to the windy conditions in the Lebanese highlands. He wrote of how "each stroke of the wind fills my garden with debris . . . this intermittent duel seems to recall Hercules striking the heads of the Lernean Hydra who fought back at each turn; in this case, it is the tree that seems strong and the wind appears to exhaust itself as it seeks to bring it to heel [*l'entamer*].[74]

Mandate-era forestry efforts did not involve such romantic concerns. State interest in development (*mise-en-valeur*) and assuring the authorities' modus operandi were preeminent. One effort focused on the fixing of shifting dunes through the introduction of *acacia saligna*. The planting of acacia was suggested because of its success in Cyprus.[75] This plant, native to Australia, had been used in French Algeria and later became "one of the worst woody invaders" in South Africa, though its impact has been less considerable in the Mediterranean.[76] Other areas, which were stable, were also considered for reforestation from a touristic point of view. This was the case for areas that were deemed uncultivable but located in regions near villages and roads for beautification. Another report claimed that forests would grow by themselves so long as the right measures were made to provide soil stability and cross-pollination. This conclusion was tied to authorities' acknowledgment that reforestation on a large scale was beyond their limited budget. Instead, reforestation focused on indigenous species and situations where there was a danger of landslide for villages, pastures, or agricultural areas. To pay for this, the local villages were expected to contribute funds, though the state budget would support money for grains, plants, and technicians.

Impact of State-Led Activity

The importance of state-led efforts as sources of ecological interventions is evident from the records of French activity in the post-Ottoman Levant. State intervention was limited by the unpredictability of ecological agency. At the same time, it is noteworthy that local administrative and public sphere elites were in otherwise rare agreement with French policies. The directionality and capacity of the French colonial state resulted in an increase in ecological transfers in the early mandate years. The longer-term impact of these transfers is difficult to discern. Early reports' optimism of the ability of the mandate state to make major changes to Syria's agricultural produce did not bear out over the course of the 1920s. Optimism about cotton development was countered by its limited growth despite efforts to encourage the spread of seeds. Cotton produced only 30,000 bales in the first year of the mandate, whereas it had stood at 180,000 bales in 1913 with German prewar estimates suggesting that Cilicia could grow up to 1 million bales.[77] Cilicia, which had been widely seen as the jewel in the Levant's agronomic crown, was lost to France in 1921, following Turkish pressure on the northern frontier.

Some successes were evident. A 1923 campaign replanting mulberry trees and improving sericulture techniques had avoided as poor a harvest as that seen in other sericulture countries. Cotton production, which had been near nonexistent in 1921, had reached 5,000 bales of 110 kilograms in 1922, 18,000 in 1923, and 25,000 in 1924.[78] Standard crops did not perform as well, dashing initial hopes for widespread agricultural improvements. Some 450,000 tons of winter cereals (wheat, barley) were produced over 650,000 hectares. Agricultural inspector Achard predicted that 3 million out of Syria's 4 million cultivable hectares could be used for winter cereal cultivation. It was estimated that 2 million tons of winter cereals would be the minimum yearly production.[79]

Yet mixed results were evidenced in a 1928 report on agriculture in the Syrian state submitted to the League of Nations. The report noted that the Aleppo *Wilāya* (district) had seen an increase in wheat growing but the other regions—Damascus, Deir ez-Zor, and Alexandretta—had seen declines in output.[80] Alfalfa was cultivated in the oasis around Damascus and in Lebanon's Biqaa valley. Trials were undertaken to bring it to the coastline and to Aleppo, where it had some success. Alfalfa was judged to be well acclimatized to the Syrian situation because of its deep roots and longevity, and its spread was encouraged because it could also provide livestock fodder.[81]

Despite early optimism, reports gradually demonstrated the difficulty in freely implementing state-directed ecological changes. A general report from 1926 admitted that certain regions of the country had experienced poor harvests because of scorching weather. However, it claimed that the peasants had remained optimistic. It also claimed this was because of governmental interventions, provision of security and infrastructure (roads, canals, irrigation), as well as agricultural experimentation and loans that had reached 375,000 Syrian pounds in the state of Aleppo. It added that the agricultural situation was being improved through the purchase of agricultural machines, research on means of fighting disease, encouragement of sericulture through the donation of 100,000 mulberry seeds, and the establishment of an education program for 1925.[82] Another 1926 report noted that agricultural progress was set back in 1925 following an invasion of parasites and locusts that had harmed 20 percent of the seeds, with certain regions of Damascus state losing half of their seeds. The report also added that the sericulture campaign had "without a doubt" been less successful than that of the preceding year, with the cocoons grown totaling around 2.7 million kilograms.[83]

Conclusion

The core premise of the mandate was the idea of tutelage by more advanced and capable nations overseeing tutees in their state-building effort. For the French in Syria, a core element of this idea was that they were making the country worthy, the *mise-en-valeur*. This represented a revamped civilizing mission no longer deeply dependent on romantic-orientalist rhetoric but shifting toward the notion of technocratic management of the territory. Ecological engineering was a keystone of this policy in a fundamentally agricultural country. This state-directed intervention in local agrarian ecologies was supported by the local political elite and public sphere in a rare display of tutor-tutee consensus that further underlined the state-building logic of agricultural interventions. While the multifarious and chaotically expressed interventions of ecological actors and conditions constrained the execution of French aims, the scope of interventions was much greater than in the less organized Ottoman-state era, thus underlying the shift from the savant and romantic-orientalist experimentation of early colonial botany and forestry to the science of agriculture and state-managed directedness and organization. This case study delineates the shift in French colonial ecological interventions over the nineteenth and twentieth centuries—from savant-romantic, individualist interactions toward more bureaucratic, technocratic, state-organized, and more immediately impactful interventions. The case study also contributes to the wider literature on ecological actors and human interventions by highlighting the need to appreciate the different scopes of impacts that human organizational groups can have in relation to ecological actors and interventions. It would seem that the hierarchical and carapace-like intervention of the modern (colonial) state in societal affairs can be said to have had a much more consequential impact on ecological interventions than earlier human-ecological interactions.

Notes

1. Edmund Burke III, "A Comparative View of French Native Policy in Morocco and Syria, 1912–1925," *Middle Eastern Studies* 9, no. 2 (May 1973): 175–86; Martin Thomas, "French Intelligence-Gathering in the Syrian Mandate, 1920–40," *Middle Eastern Studies* 38, no. 2 (2002): 1–35.

2. Richard Drayton, *Nature's Government: Science, Imperial Britain, and the "Improvement" of the World* (New Haven, CT: Yale University Press, 2000). It is notable that the very word "state" comes from the concept of estates of the sovereign, a relic of feudal organization that fundamentally attached sovereignty to agricultural holdings (79).

3. Joseph M. Hodge, "Science and Empire: An Overview of the Historical Scholarship," in *Science and Empire: Knowledge and Networks of Science across the British Empire 1800–1970*, edited by Brett M. Bennett and Joseph M. Hodge (Basingstoke, UK: Palgrave Macmillan, 2011).

4. Diana K. Davis, "Potential Forests: Degradation Narratives, Science, and Environmental Policy in Protectorate Morocco, 1912–1956," *Environmental History* 10, no. 2 (2005): 211–38. Although this is not in the purview of the present article, the forestry service in the Levant Mandate was equally present.

5. Caroline Ford, "Reforestation, Landscape Conservation, and the Anxieties of Empire in French Colonial Algeria," *American Historical Review* 113, no. 2 (2008): 341–62.

6. Ulrike Kirchberger, "German Scientists in the Indian Forest Service: A German Contribution to the Raj," *Journal of Imperial and Commonwealth History* 29 (2001): 1–26; Michael A. Osborne, "The System of Colonial Gardens and the Exploitation of French Algeria, 1830–1852," in *Proceedings of the Eighth Annual Meeting of the French Colonial Historical Society*, edited by E. P. Fitzgerald (Lanham, MD: University Press of America, 1985), 166.

7. Hodge, "Science and Empire," 13–14; Christophe Bonneuil, "Development as Experiment: Science and State Building in Late Colonial and Postcolonial Africa," *Osiris* 2 (2000): 258–81.

8. Alfred W. Crosby Jr., *The Columbian Exchange: Biological and Cultural Consequences of 1492* (Westport, CT: Praeger, 2003); Alfred W. Crosby, *Ecological Imperialism: The Biological Expansion of Europe, 900–1900* (Cambridge: Cambridge University Press, 2004); Donna Harraway, *When Species Meet* (Minneapolis: University of Min-

nesota Press, 2008); Beth Greenhough, "Where Species Meet and Mingle: Endemic Human-Virus Relations, Embodied Communication and More-Than-Human Agency at the Common Cold Unit, 1946–90," *Cultural Geographies* 19, no. 3 (2012): 281–301.

9. James C. Scott, *Seeing Like a State: How Certain Schemes to Improve the Human Condition Have Failed* (New Haven, CT: Yale University Press, 1998), 2.

10. Scott also dedicates his discussion to the role of crass materialists such as Lenin and the Bolsheviks in promoting wide-scale agricultural developments, leading to disasters such as the Lysenko experiments. See Scott, *Seeing*, 164–79, 193–222.

11. Timothy Mitchell, *Rule of Experts: Egypt, Techno-Politics, Modernity* (Berkeley: University of California Press, 2002), 252–70.

12. Jean Blottière, "L'Evolution de la Cerealiculture Algerienne," *L'Agriculture des Pays Chauds-Nouvelle Série* 2, no. 10 (April 1931).

13. M. Widiez, "La Culture du Cotonnier en Afrique du Nord," *L'Agriculture des Pays Chauds-Nouvelle Série* 2, no. 10 (April 1931): 278.

14. Marvin Harris, *The Rise of Anthropological Theory: A History of Theories of Culture* (Oxford: Altamira Press, 2001), 671–73.

15. Edmund Burke III, "Rivers, Regions & Developmentalism," in *The Environment and World History*, edited by Edmund Burke III and Kenneth Pomeranz (Berkeley: University of California Press, 2009), 84–98.

16. Edouard Achard, "Notes Agricoles Sur La Syrie" (n.d.), Ministère des Affaires Etrangères (hereafter MAE) Nantes/1SL/V/2377.

17. Stanford J. Shaw and Ezel Kural Shaw, *History of the Ottoman Empire and Modern Turkey*, vol. 2, *Reform, Revolution and Republic: The Rise of Modern Turkey* (Cambridge: Cambridge University Press, 1977), 230; Ekmeleddin İhsanoğlu, *History of the Ottoman State, Society & Civilisation*, vol. 2 (Istanbul: Research Center for Islamic History, Art and Culture, 2001), 339.

18. E. Florimond, "Agronomical Advisor to the High Commissioner to General Gouraud," Beirut (19 October 1920), MAE Nantes/1SL/V/2371. The Damascus branch oversaw smaller branches in Ḥama, Ḥoms, Al-Salamiyah, 'Omraniye, Al-Nabek, Wadi al-'Adjam, Hasbeya, Dūmā, Al-Zabadani, Beqaa, Baʿalbek, Daraa, Izra', 'Ajlūn, Salt, Al-Karak, and Al-Qaryatayn. The Aleppo branch oversaw operations in Muarra, Idlib, and Barb.

19. Achard, "Notes Agricoles Sur La Syrie" (n.d.), MAE Nantes/1SL/V/2377.

20. Huri İslamoğlu-İnan, *State and Peasant in the Ottoman Empire: Agrarian Power Relations and Regional Economic Development in Ottoman Anatolia during the Sixteenth Century* (Leiden: Brill, 1994).

21. E. Florimond, "La Reforme Foncière En Syrie," *L'Agriculture des Pays Chauds-Nouvelle Série* 2, no. 10 (April 1931): 267.

22. Achard, "Notes Agricoles Sur La Syrie" (n.d.), MAE Nantes/1SL/V/2377.

23. For an account of this service, see Elizabeth Williams, "Mapping the Cadastre, Producing the Fellah: Technologies and Discourses of Rule in French Mandate Syria and Lebanon," in *The Routledge Handbook of the History of the Middle East Mandates*, edited by Cyrus Schayegh and Andrew Arsan (London: Routledge, 2015).

24. High Commissioner's Delegate to the Syrian Federation, "Avant Propos. Economie Generale Du Programme," 68–72 (n.d.), MAE Nantes/1SL/V/2362.

25. Direction des Affaires Politiques et Commerciales, "Lettre Collective No. 109" (16 May 1926), MAE Nantes/Consulat Santiago/616PO/1/51.

26. "Historique Résumé de l'Installation et du fonctionnement des services administratifs dans les T.E.O. Nord (Cilicie), du 1er Février 1919 au 15 Février 1920" (n.d.), MAE Nantes/1SL/V/2362.

27. Etat des Alaouite, "Rapport Pour le 4eme Trimestre 1924," MAE Nantes/1SL /V/1843.

28. Sanjak of Alexandretta, "Rapport Trimestriel Avril- Mai–Juin 1924" (1924), MAE Nantes/1SL/V/1843.

29. Etat des Alaouites, "Rapport Trimestriel Janvier, Février, Mars 1924," 12, MAE Nantes/1SL/V/1843.

30. Le Mandat Au Début de 1926," 16, MAE Nantes/1SL/V/2518.

31. N.A., "Le Mandat Au Début de 1926," 16, MAE Nantes/1SL/V/2518.

32. Ministry of War, Army Chief of Staff, Section of Studies for Africa, The Orient and the Colonies, "Bulletin de Renseignements des Questions Musulmanes" (10 August 1928), MAE Nantes/1SL/V/1616.

33. Mitchell, *Rule of Experts*, 19–51; Holger Weiss, "Locust Invasions in Colonial Northern Ghana," *Working Paper on Ghana: Historical and Contemporary Studies* 3 (March 2004). Locusts had affected French administrators in Algeria, leading to an attempt to improve the country's access to fresh water. See Brock Cutler, "Imperial Thirst: Water and Colonial Administration in Algeria, 1840–1880," *Review of Middle East Studies* 44, no. 2 (Winter 2010): 168, 167–75.

34. Service des Renseignements, Poste du Liban Sud, "Etude Sommaire de la Région du Djebel Amel," MAE Nantes/1SL/V/2200.

35. Ministère des Affaires Etrangères, Direction des Affaires Politiques et Commerciales, "Lettre Collective No. 109" (16 May 1926), MAE Nantes/Consulat Santiago/616PO/1/51.

36. British Consular translation of article in el Kabaa No. 343 (8 April 1930), The National Archives, Kew, TNA/FO684/4; British Consular translation of article in Alif Ba No. 2800 (10 April 1930), TNA/FO684/4.

37. British Consular translation of article in el Qabas No. 248 (15 April 1930), TNA/FO684/4.

38. British Consular translation of article in Esh-Sha´ab No. 799 (3 March 1930), TNA/FO684/4.

39. E. C. Hole, British Consul Damascus to Fakhri al-Baroudi (4 April 1930), TNA/FO684/4.

40. International Agreement for the Establishment of an International Bureau of Intelligence on Locusts, Damascus, 20 May 1926 (London: HMG Stationery Office, 1930).

41. Geoffrey Meade, Acting British Consul in Damascus to Foreign Secretary, London (6 June 1930), TNA/FO684/4.

42. International Bureau of Intelligence on Locusts at Damascus, "General Report on the Anti-Locust (Anti-Acridienne) Campaign in Syria in 1930" (December 1930), TNA/FO684/4.

43. Rapport Economique de la Zone Aintab-Killis," MAE Nantes/1SL/V/1843.

44. Achard, "Notes Agricoles."

45. Rapport Mensuel Succinct Présenté Par le Chef des Services Economiques Pour le Mois de Juin 1922" (June 1922), MAE Nantes/1SL/V/2379.

46. Captain P. J. André, "Notes Introductives a L'Etude du Sandjak du Djebel-Bereket (Cilicie)" (n.d.), MAE Nantes/1SL/V/2362.

47. Service des Renseignements, Poste du Liban Sud, "Etude Sommaire de la Région du Djebel Amel," MAE Nantes/1SL/V/2200.

48. Francis Petit, Advisor to the Government of Greater Lebanon, "Note Pour M. le Conseiller Pour les Services Economiques" (24 November 1921), MAE Nantes/1SL/V/2379.

49. N.A., letter translated by Aziz Ghazi (n.d.), MAE Nantes/1SL/V/2436.

50. Etat du Grand Liban, Délégation Administrative, "Procès-Verbal de la Séance du 21/12/1921," MAE Nantes/1SL/V/2436.

51. Alan Mikhail, "The Middle East in Global Environmental History," in *A Companion to Global Environmental History*, edited by J. R. McNeill and Erin Stewart Mauldin (Chichester, UK: John Wiley & Sons, 2015).

52. Scott, *Seeing*, 264.

53. Etat du Grand Liban, Service de la Presse, "Journaux du 29 Octobre 1924," MAE Nantes/1SL/V/1683.

54. Etat du Grand Liban, Service de la Presse, "Journaux du 11 Février 1924" (11 February 1924), MAE Nantes/1SL/V/1843.

55. Confin de L'Euphrate, Sandjak de Deir Ez-Zor, "Rapport Trimestriel (3e Trimestre 1924)," MAE Nantes/1SL/V/1843.

56. Etat du Grand Liban, Service de la Presse, "Journaux du 9 Décembre 1924" (9 December 1924), MAE Nantes/1SL/V/1843.

57. Etat de Damas, Direction des Services Economiques, Rapport Agricole et Economique 1er Semestre de l'Anne 1923, MAE Nantes/1SL/V/1843.

58. Gouvernement D'Alep, Sanjak Autonome D'Alexandrette, "Rapport Pour le 3eme Trimestre de l'Année 1924," MAE Nantes/1SL/V/1843.

59. Ministère des Affaires Etrangères, Direction des Affaires Politiques et Commerciales, "Lettre Collective No. 109" (16 May 1926), MAE Nantes/Consulat Santiago/616PO/1/51.

60. Ministère des Affaires Etrangères, Direction des Affaires Politiques et Commerciales, "Lettre Collective No. 84" (11 October 1924), MAE Nantes/Consulat Santiago/616PO/1/51.

61. Achard, "Notes Agricoles."

62. "Lettre Collective No. 109," MAE Nantes/Consulat Santiago/616PO/1/51.

63. Etat des Alaouite, "Rapport Pour le 4eme Trimestre 1924," MAE Nantes/1SL/V/1843.

64. A. R. Conty, French Ambassador to Brazil to M. Hippeau, French Consul in Bahia (9 March 1926), MAE Nantes/Consulat Bahia/79.

65. A. R. Conty, French Ambassador to Brazil to P. Larrue, Director of the Banco do Estado de Sergipe, Aracju (9 March 1926), MAE Nantes/Consulat Bahia/79.

66. Departmento Estadual do Algodao, Estado de Sergipe to Director of the Banco do Estado de Sergipe (22 March 1926), MAE Nantes/Consulat Bahia/79.

67. M. C. Girard, Consular Agent in Maceio, to Léon Hippeau, French Consul in Bahia (27 March 1926), MAE Nantes/Consulat Bahia/79.

68. Achard, Director of Inspections at the Services Agricoles et Economiques, Ministère des Travaux Publics et de l'Agriculture, State of Syria (12 June 1925), MAE Nantes/Consulat Bahia/79.

69. Achard, "Notes Agricoles."

70. Ministère Des Affaires Etrangères, Direction Des Affaires Politiques et Commerciales, "Lettre Collective No. 84" (11 October 1924), MAE Nantes/Consulat Santiago/616PO/1/51.

71. Ibrahim Hinj, Chief of the Beni Ali Haddadin, Nadim Ismail, Chief of the Kalbiyeh, Ahmad Al-Hur, Chief of the Khayatin, Mahmoud, Chief of the Haddadin of Sahiyoun, "Programme de revendications de la secte Alaouite Auprès du General Gouraud" (19 December 1919), MAE Nantes/1SL/V/2200.

72. Etat des Alaouites, "Rapport Pour le Quatrième Trimestre 1924" (20 January 1925), MAE Nantes/1SL/V/1843.

73. Diana K. Davis, "Desert 'Wastes' of the Maghreb: Desertification Narratives in French Colonial Environmental History of North Africa," *Cultural Geographies* 11 (2004): 359–87.

74. "Les Eucalyptus," *L'Agriculture des Pays Chauds—Bulletin du Jardin Colonial et des Jardins d'Essai des Colonies Françaises* 11 (1911): 94–95, 328.

75. Pierre Montet, "Rapport de Monsieur Monnet, Inspecteur Adjoint des Eaux et Forets, Sur les Forets de la Syrie et du Liban" (n.d.), MAE Nantes/1SL/V/2200.

76. Quentin C. B. Cronk and Janice L. Fuller, *Plant Invaders: The Threat to Natural Ecosystems* (Oxford: Earthscan, 2013), 64; Ahmed Hegazy and Jonathan Lovett-Doust, *Plant Ecology in the Middle East* (Oxford: Oxford University Press, 2016), 253.

77. "Historique Résumé de l'Installation et du fonctionnement des services administratifs dans les T.E.O. Nord (Cilicie), du 1er Février 1919 au 15 Février 1920" (n.d.), MAE Nantes/1SL/V/2362.

78. "Historique Résumé," MAE Nantes/1SL/V/2362.

79. Achard, "Notes Agricoles."

80. "Renseignements Complémentaire Pour le Rapport a la Société des Nations, Anne 1928" (3 April 1928), MAE Nantes/1SL/V/1560.

81. "Renseignements Complémentaire," MAE Nantes/1SL/V/1560.

82. Ministère des Affaires Etrangères, Direction des Affaires Politiques et Commerciales, "Lettre Collective No. 109" (16 May 1926), MAE Nantes/Consulat Santiago/616PO/1/51.

83. N.A., "Le Mandat Au Début de 1926," 16, MAE Nantes/1SL/V/2518.

PART II | Institutions and Professions

Science, to Understand the Abundance of Plants and Trees

The First Ottoman Natural History Museum and Herbarium, 1836–1848

SEMIH CELIK

The concept of "Westernization" has often been used to label the period of the last two centuries of the Ottoman Empire. Most of the literature on the development and institutionalization of scientific knowledge production, dissemination, and education in the Ottoman Empire from the eighteenth century onward argues that Western institutions, methods, and curricula transferred directly to the Ottoman Empire. This implies a top-down imposition of Western values, practices, and institutions in education prior to the late nineteenth century. Developments in science and scientific knowledge production in the Ottoman territories around the middle of the nineteenth century came to be understood as the direct expansion of European values and institutions.[1]

In contradistinction to these studies, this chapter argues that the "solidification" of Ottoman scientific enterprise took place simultaneously with the developments in Europe and thus was not merely an extension of European developments. Rather than fitting a one-way diffusion model, Ottoman science must be situated within the range of European and non-European networks.[2] This chapter is a contribution to a new perspective that relocates the institutionalization of modern scientific education and research in Ottoman territories as part of a global process of integration taking place within networks of exchange. As Sebastian Conrad has suggested, historiographies that reconstructed history of knowledge through asymmetrical processes of adaption and adoption of knowledge obscured the fact that social groups in remote ends of the world "made" the Enlightenment.[3] This can be seen in the Ottoman Empire, where a wide range of institutions emerged in Istanbul throughout the first half of the nineteenth century aimed at "improving" the empire and its nature through acclimatization, crossbreeding, or simply collecting and transferring samples of fauna and flora from different climates for scientific observation.[4] The Ottoman Empire thus fits within a global history of the nascent professionalization of science.[5]

This chapter studies the first Ottoman natural history museum (*Mekteb-i Tıbbiye Numunehanesi*) and herbarium (*Botanika Bahçesi*). The museum and herbarium were established as departments of the Imperial College of Medicine (*Mekteb-i Tıbbiye-i Adliyye*), located within the compounds of Galatasarayı, in the very center of Istanbul. The museum and herbarium came to represent an interimperial space where the Ottomans highlighted local knowledge production. The original building and facilities, along with the plant, animal, and mineral collections and the adjacent herbarium burned to ashes in the fire of 1848, but the legacy of the museum and herbarium lived on well after their physical destruction.

Writing on an institution of such importance, yet one that has physically disappeared with all its rich collection of objects, poses various challenges. As one artist, Tayfun Serttaş, who created an allegorical exhibition on what he claimed to be the first Ottoman natural history museum asked, "where to start to construct the historical data when only institutional correspondence is all that is left from a museum that is erased from the official history discipline and collective memory at the present time?"[6]

Keeping in mind Serttaş's question, and based mostly on written sources, this chapter is a first attempt at a historical reconstruction of the institution, its founding, development, and demise. It aims, through an "excavation" in the archives, to outline the ideological traits, background, and functions of the museum and the network of experts, bureaucrats, and politicians around it.

A Nonexistent Museum?

Little has been written about the first natural history museum. According to Wendy Shaw, museums only became integral parts of the Ottoman imperial institutional framework after the idea of universality had waned in the late nineteenth century. For her, compared to the great European museums that gathered collections from around the world following the projects of naturalists, the Ottoman Empire focused on establishing museums that promoted a national sense of self against the potential incursion of European imperialism. This dating places the emergence of the museum almost a century after its actual emergence.

Shaw is not the only historian of the Ottomans to understand the development of museums and museology in the Ottoman territories as a "nationalistic" project and to neglect the existence of a natural history museum and herbarium right in the heart of the city. Shaw's approach points to the neglect of the earlier—and more universal—interest in natural history museums within the empire. What Shaw and others interpret as a long-standing conflict between what in the classical Ottoman Empire had been constructed as *dar-ül islam* (land of peace) and *dar-ül harb* (land of war), corresponding to today's conflicting categories of West/Christianity and East/Islam. Although actors in these two conflicting spaces interacted, their interactions were mostly based on mutual enmity, exploitation, suppression, or intrigue. An asymmetry has always dominated such relations, so much so that everything could be explained in military terms. Consequently, scientific institutions were "transferred" from one to the other based on military developments. While universality was still in play despite the conflicts, it has always developed in the hands of the polity with stronger military power, which had the capacity to dominate and colonize the others. Within that context, Ottoman museums existed in physical space, but they were understood to be built on a nationalistic idea of empire.[7]

However, before ending up with an essence of nationalistic self-definition at the end of the nineteenth century, the perception of the self and others of the Ottoman ruling elite transformed in time. During the sixteenth and seventeenth centuries, which have recently been referred to as the "Ottoman age of exploration,"[8] the Ottomans tried to make sense of the social, political, and natural landscape of different worlds, and integrate the information they gathered into a corpus of cosmological and universal knowledge. Collections of *mirabilia* and a broader literature on geography have gone through significant changes throughout these two centuries, informing how the later Ottoman Empire located itself within a globalizing map of empires.[9] Ottoman geographers and travelers of the early modern period depicted climate as a universal phenomenon. However, the place they attributed to the Ottoman Empire within that universal climate points at the distinctiveness and superiority of the climate in Ottoman territories over the rest. To put it differently, at a time when Europeans were looking for a "garden of Eden" in the new world, the Ottomans found it in their own territories.[10]

The early-modern understanding of climate, nature, and environment in "nationalistic" thinking changed in the eighteenth and nineteenth centuries. The idea of the universality and possibility of controlling nature through expertise appeared within Ottoman elite circles beginning in the 1730s. Reorienting their cultural capital around floriculture, especially of tulips, members of the Ottoman ruling elite had produced a significant amount of knowledge about the crossbreeding of different types of flowers in the eighteenth century. "Perfection" of nature through human intervention and the collection of natural objects was a prominent idea among the members of sultanic households during that period.[11] Acclimatization, crossbreeding, and transfer of fauna and flora more generally became a more systematic scientific and economic concern at the beginning of the nineteenth century. Ottoman rulers then introduced projects such as growing American cotton in Istanbul, growing Italian silkworms on Chinese mulberry trees, and breeding Merino sheep in Anatolia in order to compete with European markets in manufactures and textiles.

The first natural history books were translations from European languages into Ottoman Turkish. Mustafa Behçet Efendi translated Georges-Louis Leclerc de Buffon's *L'histoire naturelle, générale et particulière*, as *Tarih-i Tabii Tercümesi* in the first half of the nineteenth century.[12] Ottoman diplomats and state dignitaries wrote extensively about their experiences in European natural history museums and botanical gardens long before these scientific translations were published. In his *Avrupa Risalesi* (Treatise on Europe), Mustafa Sami Efendi emphasized that the botanical garden in Paris was both a place of leisure and of science.[13] Some of his information regarding the garden was based on the observations of Yirmisekiz Mehmet Çelebi, the first envoy to Paris. At the beginning of the nineteenth century, Es-Seyid Mehmed Emin Vahid Efendi mentioned the dissection laboratory (*teşrihhane*) in Pécs (in modern-day Hungary), which was depicted as a museum where body parts, fetuses, and animals exhibited in glass containers attracted the curiosity of observers.[14] Mehmet Sadık Rıfat Paşa, during his trip in Italy (ca. 1830) wrote about his observations regarding the museums and collections of plants in botanical gardens, expressing his fascination with the extensive variety of plant species on display.[15]

Interest in natural history was further developed through the publication of a textbook in Turkish on natural history in 1865. It was not only meant to attract the general public, but also added to the curriculum in secondary schools.[16] Besides personal and educational interests in natural history and natural history museums and fossil collections, only after the 1860s did the Ottoman elite start publicly debating the politics of collecting, preserving, and exhibiting. In the mid-1870s, natural history education and museums were understood to be crucial features of a "civilized" society, as demonstrated by Ali Reşit Bey, the editor of Abdullah Bey's (formerly Karl Eduard Hammerschmidt) book *Fenn-i Hayvanat-ı Tıbbiye* (Veterinary Medicine): "The fact that studying *ilm-i mevalid*, or natural history, in countries that entered the way of understanding civilization in depth is a necessity, does not need further explanation. In each [of those] countries, theories of the mentioned science are studied, and at the same time, its practices are observed in museums [*müzehane*], through which lower classes [*sıgar*] and the nobles [*kibar*] of the society benefit."[17]

Similarly, Namık Kemal, a prominent Ottoman intellectual, considered natural history as a crucial science about which Europeans from all social ranks acquired great knowledge. His idea of "progress" not only included the studying of sciences including natural history but also understanding nature through collecting objects of natural history in order to be able to control and profit from it.[18] By the end of the century, according to the most prominent scientific journal of the Ottoman Empire, *Servet-i Fünun*, natural history was among the sciences that led the civilized nations to "progress."[19]

The first natural history museum that was founded in the mid-1830s in the college complex in Galatasarayı had a strong ideological bent toward progress. Similar to its counterparts in Europe and their colonies, the Ottoman natural history museum represented the ambitions of the Ottoman ruling class to reconstruct a map of shared universal knowledge; however, it reflected the raison d'état that had been developing for almost a century, which culminated in the complex ideology of *Tanzimat*.[20] A hybrid ideology, based on the dialectic between local (read Ottoman) and global knowledge (not necessarily coming from Europe), had become the very basis of the founding of a natural history museum and herbarium in Istanbul. Contrary to the claims about the foundation of the museum as an imperialist encroachment by the West, I will try to demonstrate in the following pages that the museum and herbarium were part of the development of a global network of institutions of scientific knowledge production.

Founding a Museum in Istanbul

The actual events of the founding of the natural history museum in Istanbul remain obscure. Ottoman statesmen attempted to establish a new system of medical education that would produce new civilian medical experts for the empire throughout the 1810s and 1820s. The attempt failed due to lack of facilities and personnel, and mistakes by the administration. The Imperial College of Medicine (*Mekteb-i Tıbbiye-i Adliyye*), founded in 1836 in Galatasarayı—a palace complex dating back to the sixteenth century—hosted a somewhat more successful attempt. The school functioned both as an academy and a hospital. Many consider the Imperial College of Medicine as the beginning of medical education in Turkey, but it is not seen as the origin of natural history museums. Despite the abundance of documentation around the renovation of the old palace according to the needs of the new institution, it is nearly impossible to come across a single paragraph that explicitly mentions the museum inside the complex.

The chief physician of the period, Mustafa Behçet Efendi (1803–7, 1817–21, and 1823–34), first proposed that the Ottoman Empire needed a new health education system. He had initially planned to add the education and practice of botany and taxonomy to the curriculum. He did not mention the establishment of a museum or herbarium.[21]

Ottoman statesmen were already familiar with plant collecting and exhibiting activities. European botanists and plant collectors from Europe often visited the Ottoman territories. Requests for permission to travel and collect plant samples, especially in Anatolia, had always been welcomed by the Ottoman state, and necessary assistance had been given, including that of translators and accompanying officials.

Behçet Efendi was one of those who were aware of such activities. Although his concept of the new college did not explicitly contain plant-collecting activities and a museum, the initial building activities included storage rooms inside the palace.[22] It is unclear whether these were initially designed to be part of a future museum, but the existence of the herbarium from the early days of the college and the building work of necessary infrastructure and facilities demonstrate that the idea of opening a herbarium and museum had been conceived prior to the establishment of the college.

In 1836, an Austrian gardener, Joseph Skalak, was hired to take care of the botanical garden, in addition to two other specialist gardeners from Tarnovgrad (in modern-day Bulgaria), and nine gardener/laborers (*rençberan*) of various Ottoman origins who had already been working there. By 1838, there was constant work going on in the botanical garden and herbarium for the building of a greenhouse (*limonluk*).[23] Other account books mention expenditure on the *numunehane*, which refers to the natural history museum. Necessary facilities were built and established inside the college building before 1838. Cabinets for objects of natural history were produced in Vienna and brought to Galatasarayı; all these activities constituted more than one-third of all the expenses for the college for three months, from October to December 1838.[24]

The few sources that exist allow us to discern that the herbarium covered an area of 9,000 square meters[25] out of a total of 27,000 square meters[26] of the whole college complex. Besides the garden, two apartment buildings of unknown size were reserved for the natural history museum collection. The collection of the museum and herbarium, according to some accounts, initially contained 1,500 plant samples, mostly sent from Italy and France.[27] The collection was later enriched with plant samples from China and India collected by a Tuscan collector, Annibale Foresti, right before the fire destroyed everything. Some accounts mention plans to create a separate section of "exotic plants," without really defining what was meant by exotic. In 1844, the addition of a zoology department was decided.

The collection was further enlarged by the anatomical material that was received as a result of medical treatment, such as kidney stones and entozoa, as they were found to be "rare and odd" (*nadide ve garibe*); exhibiting such rarities was considered an important feature of natural history museums. Other "oddities" were added, like a fetus of Siamese twins and a baby sheep with six legs.[28] What constituted the main collection outside of such oddities is yet to be discovered. In the absence of a comprehensive catalogue, however, bits and pieces of information give a glimpse into the evolution of a museum that survived a little longer than a decade.

According to the inventory compiled and published as a result of an investigation of objects that survived the fire of 1848, the collection included the following: 44 examples of mounted mammals, 32 examples of mounted birds, 500 examples of mammal and bird flesh to be mounted, 54 examples of fish either mounted or protected in alcohol, 38 examples of reptiles, 1,600 examples of shellfish, 93 examples of polyps, 1 spider, 276 examples of fossils, and about 100 kilograms of minerals.[29] These damaged items were not a representative selection of the whole collection, much of which perished in the fire. We know that the museum had held a greater number of plant samples. Another unique source that gives information about the collection of the museum and herbarium were the "leftovers" from Annibale Foresti's collection of Indo-Chinese objects, which were listed as follows: 200 pieces of flower samples from China, a book of rare plants found in Indo-China, seven chests of insects, a book of paintings of various Chinese textiles and jewels, Chinese musical instruments, clothing produced in China, and so on.[30]

Plant Collectors, Doctors, Bureaucrats, and Locals

Prominent professors of botany in twentieth-century Turkey, such as Asuman Baytop and Turhan Baytop, characterized the history of the herbarium and natural history museum as one of dependency, colonization, and direct transfer of European values of science. They underestimate the agency of Ottoman statesmen and professors of the Imperial College of Medicine. Contrary to their tendency to downplay the efforts of Ottoman experts and dignitaries, Ottoman subjects were active at every level of organization of the museum and herbarium. In the words of Charles MacFarlane, a British traveler who visited the college complex in 1847–48, "no harsh criticism could apply to the liberality of the young Sultan in providing the sums necessary for stocking the establishment with implements, museums, cabinets, and other means and facilities of study."[31] While the agency and financial support of the sultan was indeed considerable, a network of scientists, bureaucrats, plant collectors, and locals made the museum experience possible. The importance of that network for the existence of the museum was celebrated by medical experts in and out of the empire, even seventeen years after the facilities were destroyed by fire in 1848.[32]

Karl Ambroise Bernard, a botanist and plant collector of Austrian origin, played a prominent role in the founding of the college and particularly the museum. Bernard had certainly set the tone for the crystallization of ideas around natural history outlined above. However, contrary to the overemphasis on his agency in the establishment of the museum and herbarium, he was not alone. Salih Efendi, first a student, later a professor of botany and the author of the first botany and zoology book in Turkish, was appointed head of the botanical garden in 1844 and was an active agent of the project. He was not only responsible for the organization of the botanical garden, but he also led missions to collect plants for the museum in the vicinity of Istanbul.[33] An equally important figure active in the museum was a certain Marko, a Greek subject of the Ottoman Empire. Later in the 1840s, Marko began teaching botany and natural history in the college and remained one of the few professors in the field throughout the 1850s.[34] In 1844, when Bernard had died, a well-known German-born Austrian botanist, Friedrich Wilhelm Noë, who was commissioned by the sultan himself to collect plant samples for the herbarium and animal fossils to enrich the newly founded zoology collection, was made the head of the museum. Not only was he commissioned by the sultan, but he was also invited to join him on his trip to Bursa in order to demonstrate his expertise. His journey to Bursa and Mount Olympus resulted in the addition of further plant and mineral samples to the natural history museum and his own personal collection. Noë was also sent to the Ottoman border with Persia, with a committee led by Derviş Pasha, who had been charged with the duty of drawing a map of the border in 1848–49.[35] Noë was dispatched and paid as a pharmacist and collector to assist Konstantin Karateodori, a professor from the college, to compile the college museum collection.[36] Furthermore, the travel book, *Seyahatname-i Hudud*, published after the negotiations and frontier setting had been completed, was written as a report, not only based on the observations of Derviş Paşa and his colleague Mehmed Hurşid Paşa but also on the information provided by Noë and Karateodori. The report gives great details on climate, natural resources, natural history, and flora of Eastern Anatolia and present-day Iraq.

Seyahatname-i Hudud provides a natural history account of the eastern provinces of the empire, together with an anthropological analysis of the periphery by the experts that represented the imperial center and its ideology. Incorporating the "non-Ottoman" agent, in this case Noë, into an imperial discourse, the book represents the Ottoman perception of the relationship between human beings and nature, which was considered as a marker of the level of civilization. The authors depicted a rural and nomadic population that viewed nature as ontologically equal to human beings and organized their everyday lives accordingly. This way of seeing nature was described as "savage" and backward. Therefore, it was an important task for the empire to discover their "nature" and translate their values into the language of "civilization."[37] Such an inquiry was also in line with the constitution of the college. It is no surprise that *Seyahatname-i Hudud* was concerned with the hygienic conditions of water resources and the physical state of dikes, and that the health of local populations and cleanliness of urban spaces were constantly reported. The information in the *Seyahatname-i Hudud* implies that the functions of the natural history museum went beyond merely providing scientific facts. It represented a hybrid space where the local and the imperial interacted.

Plants and mineral samples, together with animals, were collected within a network established between Ottoman and non-Ottoman plant collectors, doctors, bureaucrats, local officials, and local inhabitants of different provinces of the empire. In the early 1840s, the museum was well established and became the starting point for frequent expeditions into Anatolian territories for the collection of natural history objects to be brought to the museum. The Ottoman government made considerable efforts to provide assistance. In 1846, a well-prepared trip to the Anatolian provinces was organized, and the government had reserved around 15,000 guruş (around 140 pounds sterling) for a detailed list of chemicals and tools that were necessary for collecting and conserving the samples. The Ottoman government also expressed concerns about the protection of the collected objects and advised that the objects, once collected, should be sent immediately to the museum. Despite the existence of staff that had been trained in plant collection and preservation (one such example is Salih Efendi),[38] the Ottoman state wanted to rely on a team of "experts" collaborating with locals and therefore guaranteeing the best results. For that purpose, three students and a professor were chosen from the college. The whole trip was depicted and justified as an evident proof (*delil-i aleni*) of the sultan's support and love for science. It was expected to improve scientific progress in the empire, especially that of natural history and botany, and to add to the prestige of the Ottoman Empire in the world.[39]

We do not know very much about the results of the trip of 1846. However, the plant-collecting activities of Salih Efendi, the commissioning of Noë by the sultan, and the centrally organized research trips within the Ottoman Empire show that the foundation, management, and development of the herbarium and natural history museum was a result of collaboration at many different levels of state and society. The actors involved included scientists and experts of different origins, the sultan and higher-level bureaucrats, and students of the college, as well as the local population that would accompany and guide the professors in their long and difficult journeys within Ottoman territories.[40] From the early 1820s onward, plant collectors from around Europe had established good relations with locals during their trips to Anatolia. The relationship between the local population and the botanists was based on friendly collaboration, as in the experience of a ninth grade non-Muslim Ottoman student of the college, who visited his hometown in Bursa. In addition to collecting plant and mineral samples for the museum during his stay in Bursa, he wanted to vaccinate the locals who had not yet been vaccinated against smallpox.[41] Further examples can be found in *Seyahatname-i Hudud* about collaboration between the plant collectors and how local knowledge was actively utilized. While gathering information about regional climates and landscape, Noë and Karateodori often consulted locals regarding knowledge about the landscape and natural history of the respective regions.[42]

At the same time, the interest of the upper echelons of the Ottoman polity in the museum is of particular importance. Sultan Abdülmecid would personally inspect the college during the course of each academic year and attend the final examinations, where he would also ask questions of candidates as part of the examination process. During his visits to the college, he would first take some time to enjoy the herbarium and museum.[43] As explained above, Noë, who had been sent to Istanbul to collect specimens for the private collection of the king of Saxony, was appointed as the head of the herbarium and the museum of the Imperial College by Abdülmecid himself, on the advice of the king.[44] What is more, the sultan personally ordered that natural history collections be brought to the museum from Europe, through his consuls.[45]

The network of scientists, experts, and bureaucrats went far beyond those resident within Ottoman territories. The fact that there were French, Austrian, Italian, Indian, and Chinese collections in the museum can be taken as an indicator of the complexity of a network that integrated Istanbul in a broader, international network of scientific institutions. Botanists and plant collectors from various empires constantly visited the museum. In 1845, a sample catalogue of the collection was sent to St. Petersburg, extending the outreach of the network even further. In the following years, the catalogue was sent to European natural history museums on a regular basis.

The network helped Ottoman scientists and state officials to position themselves within larger European and imperial networks. Personal, economic, and political motives were at play within the relationships established around the museum. The professors or experts of non-Ottoman origins traveled to Istanbul, not only for scientific and trade purposes, but also because they tried to find employment and academic prestige. The expectation that their efforts in contributing to the herbarium and the museum would award them with a lucrative position in the college was a strong motivation.[46] Joseph Clementi, for instance, could not help but express his disappointment at not being employed as a botany and natural history professor in the college after it had been transferred to the *Humbarahane* building after the fire in 1848.[47] Scientists like Noë were collecting rare plants and fossils, not only to contribute to scientific knowledge but also because they knew that such objects were of considerable economic value. Noë used to sell sets from his discoveries in Rumelia and Anatolia to European collectors in Istanbul. Therefore, these actors were not only part of a network of experts but also of a market of goods and artifacts of natural history. In fact, some of Noë's collections of Ottoman plants had been bought in Europe before they were exhibited in the natural history museum in Galatasarayı.[48] Europe was not the only place where such goods were of particular economic and symbolic value. Although Istanbul was already the center of exchange for tulips by the first half of the eighteenth century, the natural history museum made the city part of another market: a place where plants, animals, and "objects of curiosity" were preserved from different parts of the world. Annibale Foresti's collection of Indo-Chinese and South American flora and objects became an issue of severe conflict between him and the Ottoman bureaucrats, reaching the desk of Sultan Abdulmecid himself. Although accounts written by botanical scientists and contemporaries of Foresti describe him as a man of science,[49] his own discourse around his collection, most of which turned to ashes during the fire of 1848, paints the picture of a dedicated merchant who came to the city to find a suitable market to sell his goods. And he had no shortage of customers too. Abdullah Efendi, Ethem Pasha (both members of the Ottoman Imperial Council), and the chief physician Ismail Pasha were among the names he bargained with. After a period of negotiation, he decided to sell a part of his collection to Ismail Pasha for a cheaper price than he had given to the others, since Ismail Pasha had promised him "good offices" (*en considération des bons offices qu'il m'avant promis*). However, the fire of 1848 destroyed the ambitions of the two—Ibrahim Pasha was deprived of the symbolic capital that the collection would provide him; Foresti of the reward for his troublesome journey of five years in Indochina.[50] Ottoman administrators were aware of this trade around natural objects within the empire and attempted to regulate it. The state encouraged plant and fossil collecting activities among the members of the Imperial College by buying samples that were deemed important.[51]

Although they held a certain economic value, such objects were also of symbolic and political relevance and formed part of a gift economy. Possessing or giving such objects played a symbolic role that went beyond the object's economic value. The decision of Foresti to sell a part of his collection for a lower price to someone who offered him "good services" demonstrates that fact well. On a higher political level, it is interesting to observe that, at a time of harsh competition with the Ottoman sultan, the governor of Egypt, Kavalalı Mehmed Ali Pasha, had a particular interest in the natural history museum. Mehmed Ali Pasha was the first to contribute to the zoology section of the museum with a rich collection.[52] During a visit to Egypt in 1846, Konstantin Karateodori was given further gifts of samples of plants, minerals, and animal fossils for the museum. On an interimperial level, the French emperor took Sultan Abdulmecid's order for the purchase of natural history collections with "friendly sentiments to an ally and a neighbor" (*par suite des sentiments d'amitié qui l'avinent envers son auguste allié et voisin*). Therefore, the imperial ideology, which created the grounds for the existence of an interimperial network, generated a symbolic space in which competing political actors became equals, and actors from detached parts of the world became "neighbors."[53]

Conclusion

On 12 October 1848, a fire destroyed 200 houses and 60 shops in the district of Beyoğlu. According to an account from the time, "the fire, lasting eight hours . . . destroyed the college building, pharmacy, most of the belongings of 40 professors and of almost 400 students, the herbarium, glass houses and the whole museum where the library and a physics collection were hosted. It is a loss that will never be replaced, since the museum of the college was one that was comparable to the most perfect examples in Europe."[54]

Prior to the fire, the plant collection included more than 12,000 samples of mostly rare species, according to Joseph Clementi. Although the Imperial College was destroyed together with the herbarium and the museum, the intellectual/bureaucratic infrastructure established around it did not suddenly disappear. Immediately following the disaster, Alexander von Nordmann, a Russian state advisor and head of the botanical garden in Odessa, visited the restoration site to contribute to the ongoing work. Nordmann donated a collection of fish and mineral samples and a copy of his *Fauna Pontica* to Hayrullah. The Russian traveler, botanist, and collector Pyotr Alexandrovich Chikhachyov was among those who visited the site. He had had close contact with the circle of experts and scientists around the museum in Istanbul and evidence suggests that he closely followed the aftermath of the fire.[55] The appointment of Macarlı Abdullah Bey (formerly Karl Eduard Hammerschmidt, a Hungarian refugee) as the new botany professor for the college in 1870, now hosted in the confines of *Humbarahane*, started a new institutional phase in the history of natural history museums in the Ottoman Empire, which is the subject of further research.

This chapter has demonstrated that the herbarium and natural history museum experience in Istanbul was part of a complex network that cannot merely be defined within a colonial or "Westernization" context. It reflected the nature of the nineteenth-century Ottoman Empire as part of a globalizing network of diplomats, merchants, travelers, philanthropists, and scientists. The cooperation and conflict among the corpus of experts, and between the local and the universal, had been a dominant drive for the direction of the museum and herbarium. The demise of the museum and herbarium did not destroy its significance for Ottoman history.

Notes

1. Studies conducted by Ekmeleddin İhsanoğlu in the field of history of science in the Ottoman Empire are still widely referenced. See Ekmeleddin İhsanoğlu, *Osmanlılar ve Bilim: Kaynaklar Işığında bir Keşif* (Istanbul: Nesil Yayınları, 2003). Also see Tuncay Zorlu, *Innovation and Empire in Turkey: Sultan Selim III and the Modernisation of the Ottoman Navy* (London: I.B. Tauris, 2008); Nikki R. Keddie, "Intellectuals in the Modern Middle East: A Brief Historical Consideration," *Daedalus* 101, no. 3 (Summer 1972): 39–57; Manolis Patinotis and Kostas Gavroglu, "The Sciences in Europe: Transmitting Centers and the Appropriating Peripheries," in *The Globalization of Knowledge in History*, edited by Jürgen Renn (Berlin: Edition Open Access, 2012),

321–43; Vedit İnal, "The Eighteenth and Nineteenth Century Ottoman Attempts to Catch Up with Europe," *Middle Eastern Studies* 47, no. 5 (2011): 725–56.

2. A thorough analysis of that historiography is far beyond the limits of the present chapter; however, the following studies give successful overviews of the evolution of the literature: Miri Shefer Mossensohn, *Science among the Ottomans: The Cultural Creation and Exchange of Knowledge* (Austin: University of Texas Press, 2015); M. Alper Yalçınkaya, *Learned Patriots: Debating Science, State, and Society in the Nineteenth-Century Ottoman Empire* (Chicago: University of Chicago Press, 2015).

3. Sebastian Conrad, "Enlightenment in Global History: A Historiographical Critique," *American Historical Review* 117 (2012): 1025.

4. Among others, it is possible to give the examples of attempts at growing Italian silk cocoons on Chinese mulberry trees in Istanbul, growing American cotton on a model farm, and the crossbreeding of Merino sheep in Anatolia. For discussions around the latter two, see M. Erdem Kabadayı, "The Introduction of Merino Sheep Breeding into the Ottoman Empire," in *Animals and People in the Ottoman Empire*, edited by Suraiya Faroqhi (Istanbul: Eren Yayıncılık, 2010), 153–70; Mehmet Ali Yıldırım, "Osmanlı'da İlk Çağdaş Zirai Eğitim Kurumu: Ziraat Mektebi (1847–1851)/ The First Modern Agricultural Education Institution in Ottoman: School of Agriculture (1847–1851)," *Ankara Üniversitesi Osmanlı Tarihi Araştırma ve Uygulama Merkezi Dergisi* 24 (2008): 223–40. Despite their importance, such experiences still await the attention they deserve.

5. See Darina Martykanova, *Reconstructing Ottoman Engineers: Archaeology of a Profession (1789–1914)* (Pisa, Italy: Edizioni Plus, 2010); Martykanova, "Ottoman Engineers: The Redefinition of Expert Identities during the Reign of Abdulhamid II and the Early Years of the Second Constitutional Period," *TURCICA* 45 (2014): 125–56. For the solidification of the scientific enterprise in Europe, see James McClellan III, "Scientific Institutions and the Organization of Science," in *Eighteenth-Century Science*, vol. 4 of *The Cambridge History of Science*, edited by Roy Porter (Cambridge: Cambridge University Press, 2003), 87–88.

6. The quote is taken from the catalogue for the exhibition of Serttaş's work held between 18 September and 20 November, 2015, in Studio-X/Istanbul.

7. Wendy M. K. Shaw, *Possessors and the Possessed: Museums, Archaeology, and the Visualization of History in the Late Ottoman Empire* (Berkeley: University of California Press, 2003), 220.

8. Shaw, *Possessors and the Possessed*, 94; Zainab Bahrani, Zeynep Çelik, and Edhem Eldem, eds., *Scramble for the Past: A Story of Archaeology in the Ottoman Empire, 1753–1914* (Istanbul: SALT/Garanti Kültür A.Ş., 2011).

9. Giancarlo Casale, *The Ottoman Age of Exploration* (Oxford: Oxford University Press, 2010).

10. Marinos Sariyannis, "*Aja'ib* and *Gharaib*: Ottoman Collections of *Mirabilia* and Perceptions of the Supernatural," *Der Islam* 92 (2015): 442–67.

11. For a sixteenth-century example of ideas of superiority, especially in comparison to South and East Asian climates, see Seydi Ali Reis, *Miratü'l-Memalik*, edited by Mehmet Kiremit (Ankara: Türk Dil Kurumu Yayınları, 1999), 124, 131, 158. For the geographer Aşık Mehmed's detailed accounts on the Ottoman nature and praise of the

climate in the late sixteenth century, see Aşık Mehmed, *Menazirü'l Avalim*, edited by Mahmud Ak (Ankara: Türk Tarih Kurumu, 2007), 270–73, 283, 414–21, and various others. For examples of non-European and noncolonialist "discovery of nature," see Arash Kazemi, "Across the Black Sands and the Red: Travel Writing, Nature, and the Reclamation of the Eurasian Steppe circa 1850," *International Journal of Middle Eastern Studies* 42 (2010): 591–614.

12. A handful of treatises from the period demonstrate the interests of the Ottoman elite in Constantinople on the crossbreeding and taxonomy of flowers, especially of tulips. See Ariel Salzmann, "The Age of the Tulips: Confluence and Conflict in Early Modern Consumer Culture," in *Consumption Studies and the History of the Ottoman Empire, 1550–1922*, edited by Donald Quataert (New York: SUNY, 2000), 83–107; Yıldız Demiriz, *Osmanlı Çiçek Yetiştiriciliği* (Istanbul: Yorum Sanat, 2009); Seyit Ali Kahraman, *Şükufename: Osmanlı Dönemi Çiçek Yetiştiriciliği* (Istanbuıl: İBB Kültür AŞ., 2015). Also see Elif Akçetin and Suraiya Faroqhi, "Introduction," in *Living the Good Life: Consumption in the Qing and Ottoman Empires of the Eighteenth Centuries*, edited by Elif Akçetin and Suraiya Faroqhi (Leiden: Brill, 2018), 1–38. For the interests of the Ottoman ruling elite in the early- and mid-nineteenth century in gardening and environmental design, see Deniz Türker, "'I Don't Want Orange Trees, I Want Something That Others Don't Have': Ottoman Head-Gardeners after Mahmud II," *International Journal of Islamic Architecture* 4, no. 1 (2015): 257–85. However, this interest in taxonomy and collecting natural history objects goes back to earlier times. Tülay Artan has recently discovered a rare post-mortem inventory of an Ottoman merchant from the late sixteenth century who collected objects of natural history, including animals, birds, and craws (personal communication with Tülay Artan).

13. Although the translation was completed by the 1830s, the publication of the book took three decades. Nuran Yıldırım, "Le rôle des médecins turcs dans la transmission du savoir," in *Médecins et ingéniuers ottomans à l'âge des nationalismes*, edited by Meropi Anastassiadou-Dumont (Paris: Maisonnouve&Larose, 2003), 135. Although earlier studies on natural history had existed, they were mostly geography books where natural history was a subdiscipline. See Sariyannis, "*Aja'ib* and *Gharaib*."

14. At the end of the eighteenth century, museums and botanical gardens in Europe were still viewed as spaces for the exhibition of rarities, oddities, and the "exotic," despite the interest in their scientific aspect. See Ebubekir Ratıb Efendi, *Nemçe Sefaretnamesi*, edited by Abdullah Uçman (Istanbul: Kitabevi Yayınları, 1999), 73–75, 93; Mustafa Sami Efendi, *Avrupa Risalesi* (Istanbul: Matbaatü Takvimü'l-Vekayi, 1256 [1840]), 23–24.

15. Seyid Vahid, *Fransa Sefaretnamesi* (Istanbul: Kütübhane-i Ebuzziya, 1304 [1886]), 23–24.

16. Mehmet Sadık Rıfat Paşa, *Müntehebat-ı Asar*, vol. 2 (Istanbul: Tatyos Divitçiyan Matbaası, 1290 [1873]), 16, 19–20.

17. See the introduction of Salih Efendi, *İlm-i Hayvanat ve Nebatat* (Istanbul: 1281 [1865]).

18. Abdullah Bey, *Fenn-i Hayvanat-ı Tıbbiye* (Istanbul: 1293 [1876]), 7.

19. Namık Kemal, "Maarif," *İbret* 16 (28 Rebiülahir, 1289); Kemal, "Terakki," *İbret* 45, no. 3 (Ramazan: 1289).

20. [No author], "Asr-ı Müferrit Abdülhamid-i Haniyede Terekkiyat-ı Fenniye: Müze-i Hümayun—Avrupa Matbuatı," *Serveti Funun* 49 (January 1891).

21. Tanzimat, literally "reorganization," refers to the edict of 1839, which contained principles for a program of empire-wide social and economic reforms.

22. Prime Ministry Ottoman Archives (PMO), Istanbul (PMO) C.SH 1287.

23. PMO, C.SH 534.

24. PMO, D.DRB.İ. 2–12; PMO, NFS.d 209. The fact that most of the expenses were for the carrying of the soil from different parts of the city (mostly from nearby Alemdağ) might indicate that the garden was still in preparation. PMO, C.MF 2840.

25. PMO, C.MF 656.

26. *Journal de Constantinople* 129, 24 November 1848.

27. PMO, C.MF 20-987.

28. "Mekteb-i Tıbbiye-i Şahane'nin 1870'li Yılların Başındaki Doğa Tarihi Koleksiyonu," edited and translated by Feza Günergün, *Osmanlı Bilimi Araştırmaları* XI (2009–10), 338.

29. Information on Annibale Foresti is rather scarce. His pain after losing the labor involved in his five-year-long plant-collecting expedition to the fire in 1848 is explained in detail in a petition he wrote. The response to his petition refers to him as a Tuscan plant collector. See PMO, İ.MVL 153-4358. However, details of his life need closer scrutiny and may be a subject for further research.

30. Charles MacFarlane, *Kismet or the Doom of Turkey* (London: Bosworth, 1853), 91.

31. *Gazette Medicale d'Orient* 10 (January 1866): 146–47.

32. PMO, İ.DH 4696.

33. PMO, A.MKT.UM 277-65.

34. Derviş Paşa himself was a professor at the college, teaching medical physics, geometry, and chemistry classes. He was also educated in mineralogy.

35. Asuman Baytop, "Wilhelm Noë (1798–1858) ve Türkiye Bitkileri Koleksiyonu," *Osmanlı Bilimi Araştırmaları* 13 (2012): 24; PMO, İ.HR. 49-2317; PMO, İ.DH 11981; PMO, C.SH 850. Since he was appointed after the fire that had destroyed the college and the museum in Galatasarayı, Noë was asked to directly send his findings to the *numunehane* of the college, which then was moved to the nearby barracks (*Humbarahane-i Amire Kışlası*). The mineral section of the museum seems to have moved to the building of the military college, Mekteb-i Harbiye. PMO, İ.MVL 126-3246.

36. Together with descriptions of visited cities, the culture, habits, and traditions of the inhabitants were described with comments that constructed the cultural proximity between the center and the provinces. Comments by the author(s) at times reflect the self-orientalism of the imperial elite. The two pashas thought bananas were "strangely shaped" (p. 4: *acaibü'ş şekil bir şeydir*), they commented that hunting habits of certain tribes in southeast Anatolia were "amusing" and thought that the veiled women of the region looked "ugly" (p. 11: *haricden bakanlara nisanın peçeleri fena görünürse de*). They also referred to places with unfamiliar and unfavorable climates as "scary" (p. 286: *buraları susuz ve hali mahaller olduğundan muhiş yerlerdir*). Nomadic tribes, in particular, were associated with animal-like behavior (*tavr-ı hayvani*) for their intimate life with animals and nature, and such communities were labeled as "a natural republic of disorder" (p. 25: *Cumhuriyet-i tabiyye-i gayr-ı muntazama*). For a discussion on Ottoman

orientalism, see Ussama Makdisi, "Ottoman Orientalism," *American Historical Review* 107 (June 2002): 768–96; Selim Deringil, "'They Live in a State of Nomadism and Savagery': The Late Ottoman Empire and the Post-Colonial Debate," *Comparative Studies in Society and History* 45 (April 2003): 311–42.

37. Other examples demonstrate that there were students who initiated plant-collecting activities even on their own. See PMO, A.MKT 49-22.

38. For the motivations of the trip in 1846, see PMO, İ.DH. 138-7087. For a rather rare source that lists the tools and chemicals requested for the trip, see PMO, İ.DH 139-7126.

39. Collaboration with locals can be traced in the Ottoman archival sources, such as PMO, AE.SSLM.III 65-3905; PMO, A.DVN.DVE 23A-79; PMO, A.DVN.DVE 2A-69; PMO, İ.DH 7271; PMO, HR.MKT 281-67; PMO, C.MF 3001; PMO, C.HR 4745; PMO, C.HR 5026; PMO, C.DH 11602. Contrary to the plant collectors working for the Ottoman museum or those of European countries, European archaeologists were in conflict with locals over the excavation, collection, and transportation of antiquities. Shaw, *Possessors and the Possessed*, 71–72.

40. See PMO, A.MKT. 49-22. For similar arguments in European contexts, see Richard H. Grove, *Green Imperialism: Colonial Expansion, Tropical Island Edens, and the Origin of Environmentalism, 1600–1860* (New York: Cambridge University Press, 1995); Robert E. Kohler, "History of Field Science: Trends and Prospects," in *Knowing Global Environments: New Historical Perspectives on the Field Sciences*, edited by Jeremy Vetter (London: Rutgers University Press, 2011).

41. However, one should not assume the "encounter" between the representatives of the "universal" and the "local" knowledge to be one without conflict. In fact, *Seyahatname-i Hudud* refers to local "know-it-alls" (*ukala*), whose knowledge is ridiculed in the text. One example among others can be found in *Seyahatname*, 24–25. For the importance of interaction between the indigenous people and the scientists, see Kohler, "History of Field Science: Trends and Prospects," 214–21; Grove, *Green Imperialism*.

42. *Takvim-i Vekayi* 234, 24 Ramazan 1257 [9 November 1841]. For Abdulmecid's particular interest in visiting and founding museums, see Shaw, *Possessors and Possessed*, 54.

43. *London Journal of Botany* 1 (1847): 50.

44. PMO, HR.TO 150-65.

45. Edward C. Clark, "The Ottoman Industrial Revolution," *International Journal of Middle East* 5 (1974): 70, argues that by the 1840s engineers and technicians coming from Europe earned double the wage they would earn in their home countries.

46. The response to Clementi's petition can be found in PMO, A.MKT 24-39. For Clementi's response to the denial of his employment, see PMO, HR.TO 43-36.

47. See *London Journal of Botany* 6 (1847): 50–51. As Darina Martykanova rightly points out, an important motive for Europeans to come to Ottoman centers and practice their science there was a purely economic idea of employment. Martykanova, *Reconstructing Ottoman Engineers*, 132. Sheets-Pyenson offers an analysis of five different natural history museum directors who traveled from Europe to South Africa in the mid-nineteenth century, although they had prestigious posts in their respective countries. The possibility of maximizing their income appears to have played the

most important part in their story. Susan Sheets-Pyenson, *Cathedrals of Science: The Development of Colonial Natural History Museums during the Late Nineteenth Century* (Montreal: McGill-Queen's University Press, 1988), 26–29. A similar argument was put forward by Carter Findley, who claims that the "westward reach" of Ottoman urban aesthetics and style after the eighteenth century should be understood at the same time as "eastward reach" of European artists and architects seeking employment. Carter Vaughn Findley, *Ottoman Civil Officialdom: A Social History* (Princeton, NJ: Princeton University Press, 1989), 57–58.

48. See, for example, *Annali dell'Instituto di Corrispondenza Archeologica*, vol. 37 (Rome: L'Instituto, 1865), 183, which defines Annibale Foresti as "medicus superioris ordinis in regno Turcico."

49. PMO, İ.MVL 153-4385.

50. PMO, C.MF 3001.

51. Alan Mikhail, *The Animal in Ottoman Egypt* (Oxford: Oxford University Press, 2014), 6. For the long history of imperial relations based on animal exchange/gifts between Egypt and the Ottoman center, see 112–13.

52. Alan Mikhail, *The Animal in Ottoman Egypt*.

53. PMO, HR.TO 150-65. For an argument against the perception that isolates scientific cultures from each other in conformity with a similar topology that separates human settlements on maps, see Berna Kılınç, "Ottoman Science Studies—A Review," in *Turkish Studies in the History and Philosophy of Science*, edited by Gürol Irzık and Güven Güzeldere (Dordrecht, Netherlands: Springer, 2005), 253. For similar relations established between the Ottoman sultan and Austrian emperor over the natural history museum, see PMO, İ.HR 45-2140.

54. Takvim-i Vekayi, 392, 21 Zilka'de 1264/19 October 1848, 4.

55. In September 1846, he presented one of his books to Abdülmecid, together with some engravings, most likely of Asia Minor. PMO. A.DVN.MHM 3-37; PMO. İ.HR 36-1677. He was still conducting research in Ottoman territories in 1858. See PMO. HR.MKT 236-100 for the permission and orders for assistance for his journey into Anatolia.

Inventing Colonial Agronomy

Buitenzorg and the Transition from the Western to the Eastern Model of Colonial Agriculture, 1880s–1930s

FLORIAN WAGNER

At the end of the nineteenth century, the agronomic research institute established at the Buitenzorg botanic gardens in Dutch Java transformed the way Europeans organized and conceptualized colonial agriculture. Between the 1880s and 1900, agronomists turned Buitenzorg, which had been an ordinary botanic garden for most of the nineteenth century, into an extensive agronomic research institute.[1] In laboratories and on trial fields located in the hills of West Java, they developed plants and seeds that produced higher yields and had better disease resistance. Researchers at Buitenzorg perfected crossbreeding methods and genetic plant engineering. Based on early notions of ecological engineering, they combined those methods with more general strategies of mixed planting. To play it safe, they also invented new fertilizers and pesticides. In short, Buitenzorg became the agronomic center of the tropical world, where it would encourage what can be described as a "green reform."[2] In the early nineteenth century, this tropical world was tantamount to the colonized world. Buitenzorg would grow into an icon of the colonial world that remains largely forgotten today.

Colonial experts of the early twentieth century celebrated Buitenzorg as a symbol and model for modern colonial agriculture. Auguste Chevalier, France's prolific botanist and the head of the Permanent Mission of Agriculture at the French Colonial Ministry, called it "the biggest establishment in the world for the perfecting of tropical agriculture."[3] At the German Colonial Congress in 1902, the German botanist Georg Volkens celebrated Buitenzorg for radiating "pure science."[4] Volkens was only one of some fifty German experts who had received the so-called Buitenzorg fellowship issued by the Dutch governor and the German colonial administration to train botanists and agronomists in Java. Other than those specialized experts, many European travelers stopped over at Buitenzorg when taking the grand tour of Asia. "Every globetrotter knows the botanical gardens of Buitenzorg," concluded the German Colonial Encyclopedia in 1914.[5] Indeed, Buitenzorg figured prominently in the famous *Baedeker* travel guide for India, suggesting to travelers that it was worth taking a detour to the Indonesian island of Java, just to admire the famous botanic garden. Even Japanese travel guides provided detailed maps of the experimental gardens.[6]

This chapter explores how Buitenzorg played a prominent role in the transcolonial circulation of seeds and agricultural knowledge not only in Asia, but also in Africa and in the Americas. After 1900, improved cash crops from Buitenzorg such as coffee plants, sugar cane, and paddy rice, as well as cinchona and rubber trees, thus became widely used cultivars in tropical regions throughout the world. Farmers in those regions applied Buitenzorg's agronomic methods to their crops, which were often strains bred in Dutch Java.[7] The worldwide influence of Buitenzorg has yet to be fully acknowledged, especially by historians working on tropical regions outside of Asia who are less familiar with the history of the garden.

I argue that research and examples from Buitenzorg initiated a global shift in agricultural doctrines from an older West Indian plantation model of agriculture relying on large cheap labor inputs toward an East Indian model of agronomic engineering implemented by independent indigenous farmers, ideally small-scale peasants. Unlike the West Indian/American model, which had long relied on plantations owned by Europeans and worked by non-Europeans, the East Indies' model relied on teaching autonomous indigenous peasants and villages to grow "improved" crops themselves. Starting in the nineteenth century, Buitenzorg's employees trained (and frequently forced) the colonized population to participate in professional cash crop production. In so doing, they promoted and spread a new social economy and agrarian sociology that aimed at developing colonial subjects into independent agricultural producers, instead of simply exploiting their working power on plantations.[8]

Starting in 1900, European colonial governments around the world embraced the East Indian model because of its supposed liberality, a sign of the changing nature of ideals of European colonial "trusteeship." They mostly ignored that the "Eastern model" was a hybrid model and an ideal type that hardly existed in reality. While the shift to this model was global, the transition was particularly significant for colonies in Africa that had been conquered between the 1880s and 1890s. Colonial administrations in Africa consciously imitated the East Indian model as represented by Buitenzorg because it fit their environmental, labor, and political conditions. Starting in the early twentieth century, colonial governments imported seeds, fertilizers, and agricultural techniques from Java to Africa. In the British

Empire, German East Africa, the French Congo, and the Belgian Congo, colonizers emulated Buitenzorg's agronomic laboratories, established trial fields, and trained agricultural teachers.

To be sure, the shift from plantation to small-scale farming was partly a skillful use of procolonial propaganda. Colonial administrators prided themselves on introducing an allegedly liberal and rational agricultural system that benefited both colonizers and colonized. They contrasted this "liberal" East Indian model to the "illiberal" American slave plantations. The emphasis on this binary opposition and the myth that colonizers sided against traditions of slave plantations supposedly supported the idea of a "positive" colonization in Africa and Asia. Turned into a global symbol of a rational and mutually beneficial colonialism, Buitenzorg's fame rivaled Lugard's doctrine of the "dual mandate" and even the alleged "humanitarian colonialism" proclaimed by the League of Nations.[9] Like the "dual mandate" doctrine and humanitarian arguments, colonial administrators used Buitenzorg's reputation strategically to portray colonialism as a paternalistic intervention for the benefit of the colonized people. Indeed, government officials and scientists around the world argued that Buitenzorg delivered proof that European colonization could bring about progress. Imitating Buitenzorg's scientific system expressed the desire to colonize in a modern way and to leave behind slavery and oppression. Thus, the shift from the Western to the Eastern model helped to legitimize European rule by portraying it as less racist and more scientific. The focus here is on both the real material transfers of seeds and the imagined shift that helped to legitimize colonial rule.

This study offers the first global analysis of Buitenzorg while building on existing literature.[10] A number of excellent studies have pioneered research on the garden. Recently, Robert-Jan Wille has shown how it benefited from the international scientific traditions that converged in Buitenzorg to make tropical agronomy progress.[11] Suzanne Moon emphasized Buitenzorg's role in the Netherlands' East Indies development and technological schemes.[12] Harro Maat linked Buitenzorg to the evolution of Dutch agronomic science in the metropole.[13] Eugene Cittadino used it as a case study to illustrate the emergence of Darwinian plant ecology, which provoked a paradigmatic shift toward the genetic engineering of plants.[14] In his intellectual history of the Kew Botanical Gardens near London, Richard Drayton referred to Buitenzorg, writing that it was on Dutch Java that "the 'New Botany' and the new agriculture first found a common imperial expression."[15] Ulrike Kirchberger emphasized its importance for the development of a new colonial agronomy in German East Africa.[16]

The history of Buitenzorg has yet to be integrated properly into the global history of agriculture or ecological exchange. *S'Lands Plantentuin*, as Buitenzorg was called by insiders, rarely appears in French and German colonial historiography, even though it played an important role in French and German colonies. Joseph Hodge, who emphasized the correlation between agrarian doctrines and Lugard's scheme of indirect rule and the League of Nation's trusteeship ideology, gives a prominent role to British scientific agronomy without making reference to Buitenzorg.[17] Andrew Zimmerman did not even mention Buitenzorg in his prominent study of the slavery-like sharecropping system in the United States and the transfers of its racist management model from the American South to German colonies, with a specific focus on Togo.[18] Zimmerman argues that West African colonial agriculture was primarily influenced by transfers from formerly slave-run American plantation systems to the Gulf of Guinea, while East African colonies adapted the agronomic models from the East Indies.[19]

I challenge Zimmerman's strong West-East Africa division by arguing that more colonial governments throughout all of Africa, including West Africa, looked to the East Indies in general, and Dutch Java in particular, to emulate indigenous smallholder production and efficient agronomic cultivation. Buitenzorg became a model in West Africa, along with other regions of the world, including Palestine and the Philippines.[20] The "East Indian" model steadily gained ground throughout Africa, although in the Western part, as a result of direct contact with Buitenzorg and through its global influence.

The first part of the chapter deals with the redefinition of Buitenzorg as an agronomic laboratory (whereas it was seen as a botanic garden before the 1880s) and places its emergence in the context of the ecological colonization of Dutch Java and Sumatra. The second part shows that the Dutch colonial government pursued a strategy of internationalization; for example, by awarding Buitenzorg stipends to foreign researchers. The third part analyzes how improved plants from Buitenzorg cut out similar cash crops from traditional areas of production, such as the Pará region in Brazil that had produced the famed *Hevea Brasiliensis* rubber trees. It is the final part that shows how colonial experts imitated Buitenzorg's agronomic laboratories not only in East Africa but also in West Africa, as well as in Palestine and the Philippines. In the entire world, the Buitenzorg model contributed to transforming colonial agriculture along with the social organization of work.

From Plant Collection to Genetic Engineering: The Establishment of the Buitenzorg Scientific Laboratories

What did Buitenzorg stand for, except for its name, which was the Dutch version of *Sanssouci* ("Free from worry")? The Dutch East India Company had established Buitenzorg as a hill station in Indonesia's most important island, Java, in the eighteenth century. During the British occupation (1811–15), the interim government moved Java's administrative center to Buitenzorg. After the Dutch had returned, the town continued to grow rapidly. Located some 60 kilometers south of the colony's capital Batavia, and tied to it by a railway line, Buitenzorg served as a summer residence for the Dutch colonial government and a spa town for its administrators. It was surrounded by the volcanic mountains of Western Java that secured daily rainfall and provided for moderate subtropical temperatures. This humid and equatorial climate made Buitenzorg the ideal place to grow all kinds of tropical plants and establish an experimental garden for colonial agronomy. In 1817, European botanists officially inaugurated *S'Lands Plantentuin*. A German pharmacist became its director and was assisted by a Dutch and British gardener. *S'Lands Plantentuin* would become the largest and most famous botanical garden in the tropical zone, if not the entire world. By 1910, agronomic specialists agreed that "the botanical garden of Buitenzorg is currently the most important of the world."[21] Even the future director of the Kew Botanic Gardens, Arthur W. Hill, had to admit in 1915 that Buitenzorg was "probably the most complete and extensive botanical establishment in the world."[22]

Before Buitenzorg's employees became involved in genetically engineering improved varieties of marketable cash crops, its botanists collected and catalogued more than ten thousand different tropical plants. The Buitenzorg catalogue of the world's floral heritage, begun around the 1820s, aimed at the establishment of a comprehensive database of tropical plants. Though highly impressive for the sheer quantity of collected plants and seeds, the catalogue's merit was one of passive description rather than creative improvement.

Taxonomical research was the first step toward a more interventionist ecological policy. Dutch and German botanists at Buitenzorg invented a new taxonomy for ordering plants according to a complex relational system. This taxonomy enabled them to draw comparisons between different crops and single out the most productive and resistant plants for a certain climatological and geological area. Botanists collected seeds mainly in East Asia, America, and Africa. They were brought together at Buitenzorg and then redistributed around the world to interested botanical gardens, farmers, and agricultural institutions. Nevertheless, Buitenzorg's preferred purpose in the first half of the nineteenth century seemed to be statistic registration rather than genetic engineering.

The year 1850 marked the dawn of a new era in the science of economic botany, when a Buitenzorg employee smuggled quinine out of South America's Andes regions and brought it to Java. The bark of the wild cinchona tree contained the only available medicine to reduce the symptoms of malaria. It could therefore be used for medicinal purposes and, more importantly, be commercialized. It seemed that quinine enabled European administrators and traders to remain in the tropics for extended periods without facing the constant weakness caused by malaria. That is why quinine was the first crop that was grown under the auspices of the Dutch colonial government.[23] Attempts to improve its effects led to the establishment of chemical laboratories.

A more significant turning point, however, was the appointment of directors who took Darwin's ideas of natural selection seriously, which shifted the attention to evolution and descent of plants and animals. In the view of Darwinist biology, Cittadino concluded, "physiological characteristics and hereditary issues became of prime interest."[24] While in pre-Darwinian times, plants had been mostly "improved" after the harvest, once Darwin's theories became widely known, scientists set out to modify them before they grew on the fields. The Dutch botanist Melchior Treub, who became Buitenzorg's director in 1880, was among the first to launch a program in plant physiology and genetics. Treub was responsible for the garden's modernization and internationalization. Thus, Treub added a new dimension to the scientific colonial botany that had started in the early nineteenth century with the professionalization of forestry in British India.[25]

Calling Buitenzorg's extensive gardens a botanical garden, as they were known to Europeans until the turn of the century, was certainly an understatement. The area comprised an enclosure of fifty-eight hectares (approximately 143 acres) with approximately 10,000 different species of tropical plants, an agricultural *jardin d'essai* in Tijkeumeuh of seventy-two hectares, a "virgin forest" for experimental purposes in Tjibodas (283 hectares), and a mountain garden. By the 1890s, eight laboratories conducted research in fields as varied as agricultural chemistry, pharmacology, agricultural zoology, phytopathology, physiology, and forestry. The laboratories contained the most modern equipment available, including gas lighting, water supply systems, darkrooms, and a reading room with 200 scientific periodicals.[26] The pharmacological division used the great variety of medicinal plants in the gardens to develop new drugs.[27] Papaya, for example, was well known for its healing effects among the Javanese, before Dutch specialists used its alkaloid contents to cure beriberi.[28] Experimental stations that specialized in coffee, tobacco, and rice cultivation were at the heart of the laboratories.[29] Moreover, a herbarium with 200,000 specimens, a museum, a 6,000-volume strong library, and hundreds of periodical publications—such as the multilingual *Annales du Jardin Botanique de Buitenzorg*—turned the gardens into a veritable scientific research institute.[30]

Apart from its scientific aspirations, the raison d'être of the Buitenzorg facilities was primarily colonial economy: the laboratories aimed at refining and improving those species of cash crops that had yielded the Dutch colonial state its wealth throughout the nineteenth century. It was for this purpose that Buitenzorg's botanists collected plants from all over the world—the culture garden in Tjikeumeuh, for example, cultivated all the coffee and cocoa varieties that were available globally; 300 different varieties of palm trees were planted there, too.[31] The purpose of these comprehensive collections was to select and crossbreed the plants to produce more profitable and disease-resistant crops. Subsequently, scholars added experimental improvement stations for gutta-percha, caoutchouc, tea, cardamom, vanilla, and shade-plants.[32] The professional crossbreeding of coffee and tobacco plants was used to establish new tobacco plantations in the Deli region of East Sumatra, which would go on to develop into the greenhouse of the Dutch Indies, famously known as the Deli plantation belt.[33] Financed by international capital investments, planters from all over the world grew oil palms, tobacco, tea, sisal, and rubber in the "Dollar land of Deli."

Buitenzorg supplied those planters with improved seedlings and technical know-how. Thanks to the free supply of saplings and seedlings, planters did not have to rely on a single crop but could plant several different species in order to reduce their dependency on monocultures. In this way, plantation companies survived global cash crop crises—such as, for example, the collapse of world coffee prices around 1900. Moreover, planters did not have to take the risk of planting low-quality crops, prone to infestation by diseases and parasites. Instead, they received plants that had been tested and immunized in Buitenzorg, and whose employees had supervised methods of planting and further treatment of the harvest.[34] By 1900, Buitenzorg, once a center of descriptive botany, had developed into a laboratory for applied botany. Such a laboratory, however, needed extensive financial support.

Between 1880 and 1909, the director of the Buitenzorg research station, Melchior Treub, organized a system of private funding based on a simple idea: in exchange for seeds and scientific advice, the European planters would finance the researchers at the laboratories. This was also a way that the tea planters of the Dutch Indies remunerated the head of Buitenzorg's microbiological division, who directed research on tea plants. Similarly, the General Syndicate for Sugar Production funded the commission for the cultivation of sugar. The private funding of the institute yielded benefits. For example, the sugar output per cultivated hectare was the highest in the world and relegated Hawaii to second place.[35]

Otto de Vries, director of the central rubber station in Buitenzorg, worked together with ten scientific researchers to establish experimental rubber plantations and make their output more efficient.[36] After some years of experimentation, his laboratories were able to increase the *Hevea* rubber trees' resistance to diseases and the wind by shield budding and plant breeding.[37] The *Hevea* seeds, developed into highly productive and disease-resistant plants, were exported to the whole world. In addition, Buitenzorg's chemical laboratories enhanced fertilizers and pesticides to protect plantations from insects and plant disease. This attempt was backed by methods of mixed gardening, such as "intercalary planting," which helped indigenous plantations to avoid expensive fertilizers.[38] The innovative research made Java a leader in modern colonial agriculture and anticipated the development policies of the twentieth century.

As early as 1876, researchers founded an agricultural school for the sons of Indonesian elites and Dutch administrators.[39] In several attached schools and on seven trial farms, the Javanese learned how to grow cash crops—to the benefit of the European export economy. The colonial government had reserved a budget of 109,000 florins to fund these agricultural classes, which were held in vernacular languages. During the three-year courses, indigenous and metis students not only learned how to grow cash crops, but they also received training in accounting and were instructed to use credits offered by the colonial banks. Once the graduates had returned to their fields, Buitenzorg inspectors supervised their cultivation of tea, rice, or coffee and evaluated the results. These inspectors of indigenous agriculture also toured the local residences and "informed" the peasants about the best ways to grow crops: they provided them with seed variants and cultigens, among these a dry rice version that required less irrigation and water supply.[40] The training of these *Wanderlehrer* (itinerant teachers) was inspired by European agricultural teachers of the nineteenth century who had toured the countryside to disseminate new agricultural techniques and technologies.

Professionalization through Internationalization: The Buitenzorg Stipends

At the end of the nineteenth century, Buitenzorg attracted specialists from all over the world because of its reputation as the largest laboratory of modern colonial agriculture. Commissions from France, Germany, the United States, and Russia visited the laboratory while individual researchers stayed for several months and up to two years. To satisfy their demands, Dutch authorities equipped one of the workshops for visiting researchers, offered them lodgings, a small remuneration, and free classes in the pidgin versions of Javanese and Malay.[41]

In the 1880s, the internationalization of Buitenzorg's scientific staff became part of the overall Dutch strategy to take advantage of foreign experts. European experts visited Buitenzorg upon invitation by the Dutch government in Java. German researchers, for example, reported that the colonial governor paid them a honorarium and the ferry ride on Dutch ships, and provided them with accommodation and a working space for four to five months.[42] Among foreigners, this "most liberal way" of granting access to the laboratory became so popular that Buitenzorg quickly ran out of stipends.[43]

When applications exceeded the available stipends by far, German botanists—in tandem with colonial interest groups and academies of science—lobbied the German government to establish its own Buitenzorg scholarship. In 1898, Berlin agreed to fund such a scholarship, on condition of "pursuing not only scientific but also practical purposes and letting the colonial undersecretary designate one fellow at least every fourth year."[44] All in all, some fifty Germans were nominated as official visiting fellows at Buitenzorg—on both Dutch and German grants—between the 1880s and 1914. They not only contributed to agronomic research in Java but also sent plants, seeds, and technological know-how back to Germany, "for the colonial mission of the German *Reich*." Unsurprisingly, the scholarship was often explicitly awarded to administrators from the German colonies.[45] While German researchers were always the majority among the foreigners in Buitenzorg, there were also Russian, American, Austrian, Swiss, British, French, and Belgian guest researchers, who had been invited by the Dutch.[46] Belgian agronomists went into raptures over the internationalization of Buitenzorg: "every year there is a whole regiment of scholars at the laboratories who come from all the civilized countries" and study the "excellent material" produced there.[47] The agronomic experts of the world met in Buitenzorg.

But how would the knowledge and technology developed through international cooperation in Buitenzorg spread in the world?

From Brazil to Buitenzorg: The Reorientation of East African Colonial Agronomy

Around 1900 it was not yet clear that Buitenzorg would become the center of colonial agronomy. Governments that had assumed command of the colonies in Africa and Asia some fifteen years earlier, were still in the testing phase. Faced with the task of establishing a durable cash crop economy in the new colonies, administrators had two options: either they could use local, endemic plants and grow them systematically on plantations, or they could import seeds or plants that had already been tested on professional plantations in America and Asia. Rubber production, which became particularly attractive as its price on the world market reached a peak in 1903, is a case in point. On experimental plantations in the Congo and West Africa, colonial agronomists such as the German Rudolf Schlechter initially experimented with local and endemic *Kickxia* and *Landolphia* rubber trees, but their yield proved largely disappointing. Thus, experts turned to the *Hevea* rubber trees from Brazilian plantations, whose productivity was four times higher than that of African trees. In 1899, for example, the head of the Victoria botanic gardens in German Cameroons, Paul Preuss, used a steamer of the Woermann shipping line to bring seeds of the profitable *Hevea Brasiliensis* rubber trees to Africa, along with cocoa and coffee plants. Preuss also studied planting techniques and labor policies of South American cash crop plantations and applied them in African colonies.[48] By the same token, the French Colonial Ministry sent more than thirteen expeditions to Latin America, the most famous being Eugène Poisson's mission to the renowned rubber-producing Pará region in northern Brazil. He returned to Paris with 100,000 seeds of the profitable *Hevea Brasiliensis* and 350,000 grains of the easy-to-transport *Manihot Glazivolii*.[49] Poisson's *Hevea* seeds were used in French and Belgian colonies.

Despite these impressive amounts, and in defiance of a long tradition of borrowing colonial techniques from the American settler colonies, the new colonies, most of them based in Africa, demanded a more modern scheme of colonial agriculture. Brazil, in particular, did not put itself forward as a model of up-to-date agronomy. With slavery abolished only in 1888, the country was struggling to reorganize its plantation economy. While Brazil had dominated the rubber market in the nineteenth century, the producers had missed the opportunity to update their methods of cash crop production. Botanic gardens and research stations in Brazil were few.[50] Within less than a decade, the improvement and commercialization of rubber in the East Indies would damage the Brazilian plantations so badly that they relapsed into stagnation and despair.[51] The international experts who upheld the Buitenzorg model played an important part in this process. They were responsible for transfers of "improved" knowledge, techniques, and plants from Buitenzorg to the rest of the world.

Among the long-term visitors to Buitenzorg was Franz Stuhlmann, one of the most famous German botanists, who would join the German Colonial Institute in Hamburg as an expert in colonial agriculture. Stuhlmann—together with his colleagues Walter Busse and Albrecht Zimmermann—played an important role in transferring knowledge from Buitenzorg to German East Africa.[52] In 1902, they closed down an abortive experimental station for colonial agriculture in Kwai and replaced it with a professional Biological-Agricultural Institute in Amani.[53] Situated in the Usambara Mountains, the new agricultural station was modeled on the Buitenzorg "prototype."[54] Stuhlmann and Busse had done research at Buitenzorg at the turn of the century, while Albrecht Zimmermann had been head of the Javanese coffee experimental station between 1896 and 1901.[55] Stuhlmann, whose expertise had earned him an extraordinary reputation as a German colonial official, became the first director of the Amani Institute and appointed Albrecht Zimmermann as his successor. Walter Busse stayed in Buitenzorg and organized the transfer of cash crop seeds and laboratory material from Buitenzorg to Amani. Like Buitenzorg, the Amani Institute in German East Africa was part of a hill station complex that combined agronomic trial fields, several laboratories, and the features of a spa town. Its priority was to import cultivable plants that were vital to colonial life.

In the early days of the Amani Institute in 1896, Franz Stuhlmann and Walter Busse introduced cinchona trees from Dutch Java. The free distribution of seedlings to the Javanese and large-scale plantations around Bandung had boosted the Javanese production of quinine and reduced the price of this precious remedy by 80 percent.[56] In terms of turnover, then, quinine became the most important medication in the world, with Java providing for 97 percent of world quinine production by 1930.[57]

Chemists and botanists in Buitenzorg increased the quantity of quinine in the tree's bark from 0.4 percent to 18 percent: Stuhlmann and Busse copied these methods in German East Africa, brought ever more refined species from Java, and employed Indonesian and Indian experts to guarantee their prosperity. In 1907, they had planted 25,355 trees in Amani and 66,700 trees on private plantations, which indicated the success of the project in German East Africa.[58]

At the same time, Albrecht Zimmermann was particularly keen on introducing rubber and coffee cultures to Amani. Experts in Java had imported *Hevea* rubber trees from Brazil and increased their productivity at the Buitenzorg laboratories and on trial plantations. Zimmermann hoped that professional *Hevea* plantations would soon replace the African "*Raubbau*," the harvest of caoutchouc from wild rubber trees. Dismissing the declining extensive rubber collection in the Congo, he saw the future of caoutchouc production on intensive rubber plantations as taking place in Brazil or Dutch Java. In order to introduce systematic rubber production to Africa, he imported the improved seeds from Buitenzorg to Amani.[59] His esteem for *Hevea* rubber from Java was shared by the French specialist for caoutchouc cultivation, Camille Spire, whom the French colonial minister sent on an official mission to Buitenzorg in 1901.

Spire sent plants and seeds from Buitenzorg to Paris, where they were analyzed and forwarded to the French colonies in Africa. In doing so, Spire continued a long tradition of French agricultural missions to Java, which had already led to the establishment of rubber plantations in Indochina. Moreover, young French agronomists were sent to Buitenzorg to study in the rubber laboratories. Following the Dutch example, Spire advocated the distribution of seeds among colonists, administrators, and indigenous peoples in the French colonies to encourage the cultivation of rubber trees.[60] A few years later, in 1914, the French Colonial Union dispatched the agricultural inspector of Madagascar, Fauchère, to Java, to study the cultivation of coffee at Buitenzorg. Fauchère was particularly interested in the work of his German colleague Albrecht Zimmermann, who had been Buitenzorg's expert on coffee planting for seven years and had done extensive research on parasite infestation of coffee in Java.[61]

Zimmermann's expertise derived, first and foremost, from the coffee plantations in Dutch Java, where he had been head of the *Proefstation* for coffee cultivation.[62] His advice was highly regarded in German East Africa, since early and random efforts to plant 10,000 hectares of *Coffea Arabica* and *Coffea liberica* had proven disastrous. Zimmermann arrived in 1901 from Buitenzorg and set out to combat a coffee disease that had been known in Java as *blorok*. In German East Africa, it did not have a name, because it had been hitherto unknown. Zimmermann's attempts to control the disease seemed promising. At the same time, he used grafting methods developed in Buitenzorg to replace vulnerable plants with the more resilient *Coffea canephora*, and he imported seeds of improved *Coffea Arabica* from Java.[63] Not all of these acclimatized well, but ultimately Zimmermann's attempts contributed to mastering the coffee crises of the early twentieth century, when the world market price of coffee collapsed.

Walter Busse continued to send seeds to German East Africa and emulated planting methods from Buitenzorg. He received extra money from the Amani Institute to take seed samples from Buitenzorg to Africa, among them the famous Manila banana.[64] The transfer of raw material for the colonial plantation economy went hand in hand with the employment of Javanese and Indian staff in German East Africa. They were more experienced in methods and technologies of cash crop cultivation.[65]

All in all, those transfers from Buitenzorg played a vital role in supplying the new African colonies with cash crops and medicinal plants. As a general rule, the cash crops were imported into Java, bred in the Buitenzorg laboratories or adjacent experimental stations, and reexported to other colonies. Amani imported coffee, teak, and cinchona trees; and caoutchouc, gutta-percha, bamboo, indigo, and *Erythroxylon Coca* (for the production of cocaine) from Java. As we have seen, Javanese rubber varieties replaced the species that had been imported earlier from Brazil or species native to Africa. The best quality bamboo came from Buitenzorg and was ordered in Dar-es-Salam.[66] The Javanese also produced excellent cotton seed.[67]

Although the import of cash crops such as cocoa from South America and cotton from North America and Egypt continued long into the twentieth century, it was Buitenzorg that dominated the agriculture of the "new territories" in Africa and elsewhere. Moreover, while British Ceylon and the experimental gardens in Calcutta played a minor part in exporting tea, Buitenzorg accounted for the great majority of agronomist transfers from the East Indies. Besides the imports of seeds and seedlings, well-traveled agriculturists from Germany, France, or Belgium also imitated methods developed in Buitenzorg, such as polycropping techniques (for example, importing Javanese *dadap* trees, which provided shadow for the coffee plantations), graftage, or phytopathology.[68] Stuhlmann also imported machines, among them a cotton harvester from Java.[69]

The myth of Buitenzorg's agronomic superiority prevailed, although not all of those imports proved successful. The *dadap* trees that had been planted in German East Africa fell prey to caterpillars that had not existed in Java, and not all the coffee plants from the Dutch archipelago prospered on the African continent.[70] Despite these failed acclimatization attempts—which even induced the German colonial authorities to temporarily ban imports from the East Indies—the ties between Africa and Buitenzorg remained close. Cooperation was based on the myth of the Dutch laboratory as a place of rationality and applied science. As a consequence of this myth, the laboratories themselves were emulated all over the world.

Since Amani was set up by former employees in Buitenzorg, it became a replica of the laboratory in the Dutch Indies. It redistributed its own plant variations to all the African colonies and also hosted an international community of colonial experts. Situated in the rainy Usambara Mountains, close to the Tanga-Korogwe railway, it was equipped with chemical and pharmaceutical laboratories, 250 hectares of trial fields, research stations, and a guest house for foreign visitors.[71] The general government of German East Africa only subsidized researchers who applied their research to improving commercially viable plants on white plantations and developing the "indigenous cultivation" of cash crops. Like Buitenzorg, Amani sent advisors to the local population to teach them how to plant cash crops like sisal and cotton for export to Europe.[72] Amani's directors employed expert staff from Java to make sure that the multiple agronomic techniques developed in Buitenzorg were applied adequately in East Africa.[73] Ironically, Amani, which had started as a pale imitation of Buitenzorg at the turn of the century, became its main rival in the tropical world.

Emulating the Buitenzorg Model in West Africa and the World

In West and Central Africa, colonial officers also emulated Buitenzorg. In German Cameroon, which had borrowed extensively from the West Indian model to establish cocoa plantations in the 1890s, German colonial technocrats called for a colonial laboratory based on the Buitenzorg model. As Samuel Eleazar Wendt shows in his chapter, the botanist Otto Warburg became the driving force behind a laboratory project at Cameroon's botanical garden in Victoria. Warburg was the offspring of a Jewish merchant family from Hamburg who had spent several years in Buitenzorg during the 1880s. After playing an active role in the early colonial movement in Germany, he founded the *Kolonialwirtschaftliches Komitee* (1896), an economic branch of the German Colonial Society, and he published the famous *Tropenpflanzer*, a journal for colonial agriculture. Partly financed by the German Colonial Ministry, the Kolonialwirtschaftliches Komitee became the prime mover behind the agricultural economy in the German colonies. Among its biggest successes was the dissemination of cotton and rubber plantations in German colonies, the production of palm oil with special machines, and the construction of water reservoirs in German South West Africa and railway lines in Togo and East Africa.[74]

In 1899, the Komitee urged the German government to create a "laboratory in connection with a botanical garden" in Victoria, and Warburg became its main propagator. He cited Buitenzorg as an example of a multifunctional laboratory that provided for all the needs of the colony in Cameroon: agricultural engineers would improve the fertilization of the soil, phytopathologists would combat the pests infecting tropical crops, zoologists could improve stockbreeding, pharmacologists could test new drugs, and finally, hygienists could analyze contagious epidemics and chemists could examine new elements unknown in Europe. Moreover, such a laboratory would boost the production of coffee, cocoa, caoutchouc, tobacco, spices, tannins, gums, wood, and fibrous material.[75] Warburg planned to invite foreign scholars to do research there, and, as in Buitenzorg, the directors of the surrounding coffee, tobacco, and cinchona planters would bear the costs to improve their cash crops and contribute toward the salaries of the researchers and technical staff.[76] Warburg's laboratories, and his wish to increase the cultivation of useful plants, were finally realized. Victoria became a miniature Buitenzorg, with its own chemical and botanical laboratories, trial fields, an experimental station for the clearing and reforestation of "virgin" woods, and a training school to teach practical agriculture to the colonized peasants.[77]

At the same time, the Belgian colonial administration used the Buitenzorg model to whitewash the ill-reputed cash crop production in the Congo colony. To make the international community forget the rubber scandals under Leopold II, the Belgian colonial government sent its *personnel technique agricole* to Java and British India.[78] Even before Leopold's kingship of the Congo was revoked by the Belgian government in 1908, a horticulturist trained at Buitenzorg became responsible for establishing a botanical garden in Eala, not far from the Congo River.[79]

At the height of the rubber scandals, Belgian botanists claimed to have overcome the system of forced collection of wild rubber by turning Eala into a "Congolese Buitenzorg."[80] The director of the Colonial Ministry's Agricultural Service, E. Leplae, who had visited botanic gardens in Rio de Janeiro, Paredeniya (Ceylon), Singapore, and Java, picked out Buitenzorg as the "best place for agronomists and botanists to study . . . tropical agriculture." He published a detailed report with plans and photographs of its gardens and laboratories that would provide the basis for emulation in the Congo.[81] While the *Bulletin agricole du Congo belge* left no doubt that Buitenzorg was the "most important botanic garden of the world,"[82] the administration of the Congo colony found it particularly convenient to borrow from Amani, which was based in the adjacent German colony and had already tested the Buitenzorg species on African soil. As Belgian agronomists became increasingly interested in planting the *Hevea* species on professional plantations instead of forcing the Congolese to collect wild rubber, they imported "improved" rubber trees from Amani. The colonial government at Léopoldville also showed great interest in the successful sisal production launched in German East Africa. Among other plants, the Belgian administration imported coconut palms, coffee, cocoa, and bamboo directly from Buitenzorg. It even emulated the Javanese irrigation techniques for Congolese rice paddies.[83] One important reason why Leopoldville turned to the Buitenzorg-Amani model of a professional and rational botanic economy was to put the rubber scandals behind them. Another reason was that they simply wanted to make Congolese agriculture more efficient and competitive, ironically by forcing the Congolese to cultivate new crops from the East.

The connections between Buitenzorg and the botanic gardens of the British Empire had always been exceptionally close. British colonial experts even considered the Javanese botanic garden as a genuinely British imperial achievement. They acknowledged Buitenzorg's distinction ungrudgingly and hinted at the fact that a botanist from Kew, James Hooper, had helped establish Buitenzorg in 1817. Throughout the ensuing years of the nineteenth century, the Royal Botanical Gardens in Kew and Buitenzorg frequently exchanged plants and expertise.[84] At the end of the nineteenth century, however, the British realized that their own botanic gardens lagged far behind Java's sophisticated agronomic economy.

The British botanist William Thiselton-Dyer, who was director of the Royal Botanical Gardens in Kew between 1885 and 1905, was a driving force behind the emulation of the Buitenzorg model in the British Empire. He was among the earliest scientists to realize the importance of Buitenzorg and reordered the system of British botanical gardens according to the necessities of the colonial administrations and the "new botany" that served colonial purposes. He was proactive in equipping the gardens in Calcutta, Ceylon, Mauritius, and Jamaica with herbaria, laboratories, and libraries. These gardens became centers of economic botany that managed smaller establishments throughout the countryside to spread smallholder cash crop production.[85]

The research station in Aburi (Gold Coast), in particular, followed this model and sent traveling instructors to the Ghanaian cocoa planters. The latter produced cocoa on smallholder plantations of five to six acres and pushed the overall production of Gold Coast cocoa from 2,000 tons in 1902 to 40,000 tons in 1911. Starting in the 1880s, Ghanaian farmers had imported cocoa plants from São Tomé in the Gulf of Guinea, an island that was known to be modeled on the West Indian slave-plantation type. During the First World War, however, the São Tomé plants fell prey to fungal infestation by *Thrips* insects. It seems like deficient planting techniques led to the decline of cocoa plantations in British West Africa.[86] In the meantime, experts from Buitenzorg, such as the director of the Institute for Plant Disease C. J. J. van Hall, had closely observed and studied the cocoa smallholders in Ghana. Based on his global research on the improvement of cocoa planting, Van Hall advised them on how to combat diseases caused by *Thrips* and use budding methods developed at Buitenzorg. Also, in Southern Nigeria, officials such as the Assistant-Superintendent of Agriculture Frank Evans helped to proofread, publish, and spread Hall's tutorials in English.[87]

In French West Africa, the colonial administration established similar agronomic institutes and training schools for the African populations. Auguste Chevalier, who was the *spiritus rector* of French colonial agronomy and edited the most important journal of colonial agriculture in France, reorganized the colonial agronomy of French West Africa and French Central Africa. Chevalier had long fought against the French Colonial Ministry's habit of granting large-scale concessions to European companies. Instead, he favored the participation of indigenous peasants in developing a stable colonial economy. By assigning Chevalier the task of reorganizing colonial agriculture as the head of the Mission for Studying the Colonial Cultures and Experimental Gardens in 1912, the French Colonial Ministry conceded victory to Chevalier's model of indigenous smallholder entrepreneurs. To teach them efficient and modern planting techniques, Chevalier set out to establish specialized research stations for the cultivation of cotton, cocoa, coffee, and oleaginous plants. In search of a model for these stations, Chevalier traveled to the British and Dutch Indies. On Chevalier's initiative, the French colonial administration emulated Buitenzorg's laboratories and copied its scientific journals on tropical agriculture to spread a new version of agriculture.[88] Before returning to Europe, Chevalier established research institutes for tea, *Hevea* rubber, and coffee in Indochina, which was closer to Buitenzorg and more suitable for preparing plants for use in Indochina and Africa. In Senegal, he created a research station for groundnut production, a cash crop grown by Senegalese farmers that provided half of the exports of French

West Africa.[89] Chevalier's initiative was the prelude to the reform of French colonial agronomy in West Africa.

Ultimately, Buitenzorg became a model for the entire world. Otto Warburg used his botanical initiation at Buitenzorg and experience as head of the *Kolonialwirtschaftliches Komitee* to take colonial projects to a new level. He joined the Zionist movement and became the third president of the World Zionist Organization in 1911. His commitment to Zionist colonization in Palestine aligned with the practicalist branch, which favored the creation of agricultural colonies to support a Jewish homeland in Palestine. While still editing the *Tropenpflanzer* and directing several colonial cash crop companies in German Togo and Cameroon, he became the head of the Zionist Commission for the Exploration of Palestine from 1903 to 1907 and figured prominently in the Palestine Land Development Company (1908), which acquired territory in Palestine and trained Jewish settlers in farming and agronomy.[90] Warburg frequently evoked similar plans of the U.S. government to establish a new "Buitenzorg" in its colonies in Cuba, Costa Rica, and the Philippines, all acquired in 1898. He explicitly referred to the famous Taft Commission, which reorganized the administration of the new possessions.

The U.S. Taft Commission—which equipped the Philippines with a civil administration between 1900 and 1901—had created an Insular Bureau of Agriculture in Manila, which dispatched the botanist Elmer Merrill to Buitenzorg in 1902.[91] Merill's greatest interest was in forestry. In Buitenzorg, he learned how to determine the rate of tree growth and evaluate a tree's economic importance. He copied the Dutch forestry regulations and brought a map showing the method of charting *djati*, teak forests used for ship building and as railway sleepers. This method would soon be applied in the Philippines as well. Finally, he carefully took photographs of every detail in the Buitenzorg laboratories and put the Philippine Forestry Bureau on the "permanent mailing list" of the *S'Lands Plantentuin* to receive its comprehensive publications.[92] In the following years, the botanical infrastructure in Manila was turned into a replica of the Buitenzorg system, and the laboratories provided the basis for the globally admired system of colonial technocracy in the Philippines. As Warwick Anderson put it, the Philippine laboratories became the "locus of colonial modernity."[93] Merrill, instead, turned into the U.S. doyen of botanical research by editing several journals and directing the Science Bureau in the Philippines and the botanical gardens in New York. He supervised the botanical collection of Harvard University and created the Biological Laboratory and Botanical Garden in Cuba.[94]

Conclusion

The Buitenzorg model spread around the world. It triggered a green reform in Africa and Asia and influenced colonial agriculture in most of the so-called new colonies, where colonial administrators pretended to have learned from the errors of the past and tried to establish a more modern form of colonial agriculture. The main element of this policy was the portrayal of agricultural colonization as rational and cooperative. Colonizers prided themselves in gradually abolishing the plantation system that derived from the American or "West Indian" model of slave plantations.

Self-styled progressive colonial experts or institutions (such as Stuhlmann, Thisleton-Dyer, Warburg, Chevalier, and Merrill) were not entirely wrong when they dismissed the American model, which was at pains to emancipate itself from its origins in slavery. But they were wrong in portraying the East Indian model as a system of agricultural production that was completely free of coercion and forced labor. They drew a sharp line between the American slave-plantation model and the East Asian model that turned the native population into independent entrepreneurs. However, this binary opposition never existed. The Dutch model derived from the period of the cultivation system (1830s–70s) during which the Netherlands forced Javanese village communities to grow cash crops that the Dutch government would buy at a minimum price. This system had only been abolished in certain fields. Moreover, a new plantation economy emerged in Sumatra. Here, forced labor and slavery-like conditions turned the plantations into internment camps that deprived the indentured laborers of their mobility, property, and occasionally identity by retaining their identification cards.[95]

In many respects, Buitenzorg's progressive role was imagined rather than real. For most of the colonized Indonesians and indentured laborers from all over Asia, who were forced to grow cash crops for Europeans, it was even regressive. They found themselves in slave-like conditions on the plantations in Sumatra or were coerced to grow crops that they did not consume.

Nevertheless, the imagined progress became a reality in various fields. The desire to implement a new economic sociology that outsourced colonial productivity to the native population was real. After all, it was cheaper to let the colonized produce export crops than to import and protect settlers or planters. Buitenzorg was a semiofficial institution and not always in line with the policy of the Dutch colonial government. On various occasions, its founding father Melchior Treub entered into conflict with the Dutch governor.[96] Therefore, Buitenzorg did not represent colonial policy as such.

The laboratories and the botanical gardens were multifunctional and should serve the purposes of scientific progress and colonial productivity alike.

Its full meaning is exposed by its international impact. It is exactly because Buitenzorg represented the myths of rationality, profitability, and scientific progress combined that its emanative power was unbreakable for decades. According to contemporary observers, Buitenzorg even dwarfed the achievements of British colonial agronomy symbolized by the Kew Botanic Gardens. When the German Stuhlmann traveled to British India to study the production of indigo, he even bemoaned the backwardness of British colonial agronomy: "Despite the many huge plantations in British India, the methods of cultivation have not changed in the last thousand years. The British want to make quick money and do not rationally plan for the future, as long as they earn money. Only now, as they are in a crisis, they have employed two chemists (for the entire colony!), while Java has a scientific institute that has been improving indigo cultivation for several years now."[97]

It was Buitenzorg's global reputation rather than its actual achievements that guaranteed its success. Whatever the final judgment about the consequences for the colonized populations might be, this reputation served as an alleged proof that colonization can be "positive" and "rational." Consequently, Buitenzorg has to be assigned a role in colonial history that equals Lugard's policy of the "dual mandate" and the League of Nation's claim to a "humanitarian" colonialism in importance. Buitenzorg's impact could not only be felt among the colonial theorists in Europe but also among the inhabitants of the colonies. The idea that Buitenzorg was not a "political" institution has apparently led historians to exclude it from colonial history. But like Lugard's "dual mandate" and the League of Nations "humanitarian colonialism," Buitenzorg helped to justify, legitimize, and perpetuate colonial rule. Even more so, it changed the lives of the colonized populations on the ground, sometimes for the better but often for the worse.

Notes

1. The progressive role of Buitenzorg was based on the initiative of agronomist Melchior Treub, who joined forces with interested planters. Only in 1905 did the colonial government of the Dutch East Indies create a Department of Agriculture in order to increase its control over food production: Melchior Treub, *Schematische Nota over de Oprichting van een Agricultur-Department in Nederlandsch-Indie* (Buitenzorg, 1902); F. A. F. C. Went, "Melchior Treub," *Mannen en Vrowen van Betekins* 41 (1911): 11; see also Florian Wagner, "From the Western to the Eastern Model of Cash Crop Production: Colonial Agronomy and the Global Influence of Dutch Java's Buitenzorg Laboratories, 1880s–1930s," in *Agrarian Reform and Resistance in an Age of Globalisation: The*

Euro-American World and Beyond, 1780–1914, edited by Joe Regan and Cathal Smith (New York: Routledge, 2019).

2. Though not yet a "green revolution," as described by Jonathan Harwood, *Europe's Green Revolution and Others Since: The Rise and Fall of Peasant-Friendly Plant Breeding* (London: Routledge Taylor & Francis Group, 2011), 115. See also Joachim Radkau, *The Age of Ecology* (New York: Wiley, 2014); Christopher Cumo, *Plants and People: Origin and Development of Human–Plant Science Relationships* (London: CRC Press, 2015), 197–216; R. E. Evenson and D. Gollin, "Assessing the Impact of the Green Revolution, 1960 to 2000," *Science* 300, no. 5620 (2003): 758–62.

3. Letter written by Auguste Chevalier to G. Angoulvant from 24 June 1922, cited in Gabriel Angoulvant, *Les Indes néerlandaises leur rôle dans l'économie internationale* (Paris: Le Monde Nouveau, 1926), 748.

4. G. Volkens, "Der botanische Garten zu Buitenzorg und seine Bedeutung für den Plantagenbau auf Java und Sumatra," in *Verhandlungen des Deutschen Kolonialkongresses 1902* (Berlin: 1902), 182–93.

5. "Buitenzorg," in *Deutsches Koloniallexikon*, vol. 1 (Leipzig, Germany: Quelle & Meyer, 1920), 250–51.

6. See map "Buitenzorg," in *Indien: Handbuch für Reisende*, edited by Karl Baedeker (Leipzig, Germany: Baedeker 1914), http://www.lib.utexas.edu/maps/historical/baedeker_indien_1914/txu-pclmaps-buitenzorg_1914.jpg; Imperial Japanese Government Railways, ed., *An Official Guide to Eastern Asia Trans-Continental Connections between Europe and Asia: East Indies* (Tokyo: The Imperial Japanese Government Railways, 1917).

7. For the best overview, see Suzanne Moon, *Technology and Ethical Idealism: A History of Development in the Netherlands East Indies* (Leiden: CNWS Publications, 2007) and Harro Maat, *Science Cultivating Practice: A History of Agricultural Science in the Netherlands and Its Colonies, 1863–1986* (Dordrecht, Netherlands: Kluwer Academic Publishers, 2001). Eugene Cittadino and Richard Drayton mention its importance: Eugene Cittadino, *Nature as the Laboratory: Darwinian Plant Ecology in the German Empire, 1880–1900* (Cambridge: Cambridge University Press, 1990), 134–43; Richard Drayton, *Nature's Government: Science, Imperial Britain, and the "Improvement" of the World* (New Haven, CT: Yale University Press, 2000), 247–48.

8. Tellingly, the origins of the East Indian model could be found in the violent era of forced cultivation (*Cultuurstelsel*) in Dutch Java (1830–70), when Javanese farmers had to grow cash crops for the Dutch colonial government. It was remodeled as a strategy of indirect rule in the 1870s and said to be more liberal than the former slave plantations in the Americas, although it could result in similar patterns of forced labor: S. Alatas, *Democracy and Authoritarianism in Indonesia and Malaysia: The Rise of the Post-Colonial State* (London: Palgrave Macmillan, 1997), 57.

9. Susan Pedersen, *The Guardians: The League of Nations and the Crisis of Empire* (Oxford: Oxford University Press, 2015), 29, 93, 107–11, 299; Antony Anghie, *Colonialism, Sovereignty, and the Making of International Law* (Cambridge: Cambridge University Press, 2004), 115–95, 252–53.

10. Richard Drayton briefly remarks that Buitenzorg influenced the agronomic policy of the British Empire: Drayton, *Nature's Government*, 248.

11. Robert-Jan Wille, *Mannen van de microscoop: Laboratoriumbiologie op veldtocht in Nederland en Indie, 1840–1910* (Nijmegen, Netherlands: Vantilt, 2019).

12. Moon, *Technology and Ethical Idealism*.

13. Maat, *Science Cultivating Practice*.

14. Cittadino, *Nature as the Laboratory*, 134–43.

15. Drayton, *Nature's Government*, 247.

16. Ulrike Kirchberger, "Wie entsteht eine imperiale Infrastruktur? Zum Aufbau der Naturschutzbürokratie in Deutsch-Ostafrika," *Historische Zeitschrift* 271 (2010): 62–63.

17. J. M. Hodge, *Triumph of the Expert: Agrarian Doctrines of Development and the Legacies of British Colonialism* (Athens: Ohio University Press, 2007), 90–143.

18. Andrew Zimmerman, *Alabama in Africa: Booker T. Washington, the German Empire, and the Globalization of the New South* (Princeton, NJ: Princeton University Press, 2010), 16. Arguably, Buitenzorg only played a minor role in cotton growing. But the history of colonial agronomy in general cannot be written without Buitenzorg.

19. Andrew Zimmerman, "Ruling Africa: Science as Sovereignty in the German Colonial Empire and Its Aftermath," in *German Colonialism in a Global Age*, edited by Bradley Naranch and Geoff Eley (Durham, NC: Duke University Press, 2014).

20. The development of a politically influential elite of African cocoa producers in Ghana, Ivory Coast, and Cameroon substantiates this claim; see, for example, Elizabeth Wrangham, *Ghana during the First World War: The Colonial Administration of Sir Hugh Clifford* (Durham, NC: Carolina Academic Press, 2013).

21. "Le Dr. Treub," *Bulletin agricole du Congo belge et du Ruanda-Urundi* 1 (1910): 320.

22. Arthur W. Hill, "The History and Functions of Botanic Gardens," *Annals of the Missouri Botanical Garden* 2, no. 1–2 (1915): 210–11; see also Andreas Zangger, *Koloniale Schweiz: Ein Stück Globalgeschichte zwischen Europa und Südostasien (1860–1930)* (Bielefeld, Germany, 2014), 384–85.

23. Lucille Brockway, *Science and Colonial Expansion: The Role of the British Royal Botanic Gardens* (New York: Academic Press 1979), 103–9; Daniel R. Headrick, *The Tentacles of Progress: Technology Transfer in the Age of Imperialism, 1850–1940* (Oxford: Oxford University Press, 1988), 231–37; both cited in Maat, *Science Cultivating Practice*, 39.

24. Cittadino, as summarized in Maat, *Science Cultivating Practice*, 41.

25. Several authors have emphasized the transnational dimension of colonial forestry in the nineteenth century: Brett M. Bennett, *Plantations and Protected Areas: A Global History of Forest Management* (Cambridge, MA: MIT Press, 2016); Ulrike Kirchberger, "German Scientists in the Indian Forest Service: A German Contribution to the Raj?," *Journal of Imperial and Commonwealth History* 29 (2001); Richard H. Grove, *Green Imperialism: Colonial Expansion, Tropical Island Edens and the Origins of Environmentalism* (Cambridge: Cambridge University Press, 1995), 309–79; S. Ravi Rajan, *Modernizing Nature: Forestry and Imperial Eco-Development, 1800–1950* (New York: Oxford University Press, 2006).

26. G. Volkens, "Der botanische Garten zu Buitenzorg und seine Bedeutung für den Plantagenbau auf Java und Sumatra," in *Verhandlungen des Deutschen Kolonialkongresses 1902* (Berlin: 1902), 189.

27. For the experiments with fertilizers, see P. J. S. Cramer, "La culture de la Patate à Java," *Revue de botanique appliquée et d'agriculture coloniale* 20, no. 3 (1923): 237.

28. "Mittheilungen aus dem chemisch-pharmakologischen Laboratorium des Botanischen Gartens zu Buitenzorg (Java)," *Berichte der deutschen chemischen Gesellschaft* 23, no. 2 (1890). In 1929, the Dutch Christiaan Eijkman was awarded the Nobel Prize for his efforts to cure beriberi.

29. Joseph Chailley, *Java et ses habitants* (Paris, 1914), xlii–xliii.

30. Elmer Merrill, *Report on Investigations made in Java in the Year 1902 to the Department of the Interior, Forestry Bureau* (Manila, Philippines: Bureau of Public Painting, 1903), 75; Chailley, *Java et ses habitants* (1914), xlv.

31. Volkens, "Der botanische Garten," 184.

32. Volkens, "Der botanische Garten," 185.

33. Chailley, *Java et ses habitants* (1914), xlviii–xliv; Ann Laura Stoler, *Capitalism and Confrontation in Sumatra's Plantation Belt, 1870–1979* (New Haven, CT: Yale University, 1995), 14–15.

34. Volkens, "Der botanische Garten," 186.

35. Jan J. van Klaveren, *The Dutch Colonial System in the East Indies* (The Hague, J. J. Van Klaveren, 1953), 182.

36. Otto de Vries, *Estate Rubber: Its Preparation, Properties and Testing* (Batavia: Drukkerijen Ruygrok, 1920), 445; Herbert Wright, *Hevea Brasiliensis or Para Rubber* (London: 1912); P. van Romburgh, *Les Plantes à Caoutchouc et à Gutta-Percha cultivés aux Indes Néerlandaises: Des diverses espèces. Leur culture rationelle* (Batavia: G. Kolff, 1903); see Warren Dean, *Brazil and the Struggle for Rubber: A Study in Environmental History* (Cambridge: Cambridge University Press, 1987), 63.

37. Ch. Coster, "The Work of the West Java Research Institute in Buitenzorg," in *Science and Scientists in the Netherlands Indies*, edited by Pieter Honig and Frans Verdoorn (New York: Stechert, 1945).

38. Intercalary planting combined different plants in one field, which increased the resistance of both specimens to parasite infestation. Chailley, *Java et ses habitants*, xliv.

39. Maat, *Science Cultivating Practice*, 41.

40. Chailley, *Java et ses habitants*, xliv.

41. Chailley, *Java et ses habitants*, xlix.

42. Bundesarchiv Berlin-Lichterfelde (BArch), R 1001/8604, Nr. 44/45 Letter Kaiserlich-deutsches Generalkonsulat für Niederländisch-Indien to Kolonialabteilung, 29 April 1898.

43. BArch, R 1001/8604, Akademie der Wissenschaften to Foreign Ministry, 24 August 1897; Abschrift 13 December 1895, Engler and Schwendener to Minister für Geistliches, Unterricht und Medizinal-Angelegenheiten.

44. BArch, R 1001/8604, Nr. 45/56, Minister des Inneren und Reichsschatzamt, 31 May 1898.

45. BArch, R 1001/8604, Nr.78-80, "Verzeichnis der aus Java mitgebrachten Nutzpflanzen (von Giesenhagen)" and Nr. 78-85 79/78-83-85: Letter Giesenhagen to Hohenlohe Schillingfürst (Foreign Ministry), 27 April 1900. Beginning in 1911, Germans could use the Buitenzorg scholarship to visit other agronomic institutions; for example, in British India: R 1001/8605, Nr. 4-5 Reichskolonialamt to Gouverneur in Daressalam,

3 April 1911. For the significance of the Buitenzorg scholarship in German colonial botany, also see Samuel Eleazar Wendt's chapter in this volume.

46. K. W. Dammermann, "The Quinquagenary of the Foreigners' Laboratory at Buitenzorg, 1884–1934," *Annales du Jardin Botanique de Buitenzorg* 45 (1935).

47. "Le Dr. Treub," 320.

48. Paul Preuss, *Expedition nach Central- und Südamerika 1899/1900* (Berlin: Verl. des Kolonialwirtschaftlichen Komittees, 1901).

49. ANOM FM AFFPOL MIS 76bis, Mission gratuite d'études des arbres à caoutchouc au Brésil et à la Trinité d'Eugène Poisson (1898/1901).

50. Hill, "The History and Functions of Botanic Gardens," 218.

51. Dean, *Brazil and the Struggle*, 5.

52. For more information on Walter Busse, see Samuel Eleazar Wendt's chapter in this volume.

53. Hermann Paasche, ed., *Deutsch-Ostafrika: Wirtschaftl. Studien*, 2nd ed. (Hamburg, Germany: Süd-West-Verl, 1913), 246–47.

54. BArch, R 1001/8604, Direktor der botanischen Zentralstelle für die Kolonien in den Königlichen Botanischen Garten zu Berlin, Adolf Engler, to the Kolonialabtheilung im Auswärtigen Amt, 30 November 1906.

55. See Albrecht Zimmermann, *Der botanische Garten zu Buitenzorg auf Java* (Berlin: 1899).

56. Roy Porter, *The Greatest Benefit to Mankind: A Medical History of Humanity from Antiquity to the Present* (London: HarperCollins, 1997), 465–66.

57. The best-selling remedy quinine was nevertheless also used as a raw material in industrial production: Friedrich August Flueckinger, *The Chinona Barks* (Philadelphia: Blakiston, 1884), iii; Erni, "Die Krankenfürsorge in Niederländisch-Indien," *Archiv für Schiffs- und Tropenhygiene* 3, no. 3 (1899): 145–46.

58. Franz Stuhlmann, *Beiträge zur Kulturgeschichte von Ostafrika* (Berlin: 1909), 434–40; about Dutch achievements, see also M. H. Sambuc, "Le developement économique de l'Indochine et la culture du riz," *Quinzaine Coloniale*, 25 April 1910, 288–89.

59. Albrecht Zimmermann, *Der Manihot-Kautschuk: Seine Kultur, Gewinnung und Präparation* (Jena, Germany: 1913), 311.

60. FR ANOM 50COL 14: Mission à Java de Camille Spire, médecin des Colonies, pour l'étude du jardin botanique et du laboratoire de Buitenzorg, prolongation de la mission: Rapport au ministre des Colonies (copie) (1901). See also Camille Spire and André Spire, *Le Caoutchouc en Indochine: Étude Botanique Industrielle et Commerciale* (Paris: 1906), 200–209, 225, 233–34. Spire's brother André was the general secretary of the association of caoutchouc importers in France; see also Camille Spire, "Java: La Culture des Arbres Fruitiers à Java," *Revue Coloniale (Ministère des Colonies, Service géographique des missions)* 7, no. 4 (1901): 208–15.

61. FR ANOM 50COL68, Mission d'études sur la culture du café aux Indes néerlandaises par Fauchère, inspecteur d'agriculture à Madagascar, à l'initiative de l'Union coloniale française (1914); A. *Fauchère, Culture Pratique du Caféier et Préparation du Café* (Paris: Challamel, 1908), 168–69.

62. See Zimmermann, *Der botanische Garten*; J. C. Koningsberger and Albrecht Zimmermann, *De dierlijke vijanden der koffiecultur op Java*, Mededeelingen uit's Lands

Plantentuin (Batavia, Dutch East Indies: Kolff, 1901); Albrecht Zimmermann, *Kaffee* (Hamburg, Germany: 1926).

63. Albrecht Zimmermann, "Erster Jahresbericht des Kaiserlichen Biologisch-Landwirtschaftlichen Instituts Amani," *Berichte über die Land- und Forstwirtschaft in Deutsch-Ostafrika* 1, no. 6 (1903): 441.

64. Zimmermann, "Erster Jahresbericht," 440.

65. Zimmermann, "Erster Jahresbericht," 435.

66. "Wirtschaftliche Ergebnisse der Reise des Herzogs Johann Albrecht zu Mecklenburg nach Java," *Deutsches Kolonialblatt* 22 (1911): 87.

67. Reichskolonialamt, ed., *Der Baumwollanbau in den deutschen Schutzgebieten, Seine Entwicklung seit dem Jahre 1910* (Jena, Germany: Fischer, 1914), 71.

68. Also the Congo Free State imported seeds from Buitenzorg: "Rapport, presenté aux Chambres par le Ministre des Colonies du 14 Septembre 1910," *Bulletin Agricole du Congo Belge* 1, no. 1 (November 1910): 16.

69. Stuhlmann, *Beiträge zur Kulturgeschichte*, 487.

70. Albrecht Zimmermann, "Zweiter Jahresbericht des Kaiserlichen Biologisch-Landwirtschaftlichen Instituts Amani," *Berichte über die Land- und Forstwirtschaft in Deutsch-Ostafrika* 2 (1904): 214–15.

71. Zimmermann, "Zweiter Jahresbericht," 206–7; BArch, R 1001/8604, 155/157: Innenministerium to Reichskolonialamt, 22 July 1908.

72. A. Engler, "Das biologisch-landwirtschaftliche Institut zu Amani in Ost-Usambara" *Notizblatt des Königlichen botanischen Gartens und Museums zu Berlin* 4, no. 31 (1903): 63–66; see Kenneth Ingham, "Deutsch-Ostafrika: Ein wirtschaftliches Experiment in Afrika" *Afrika-Verein* 3, no. 1 (1961).

73. Kirchberger, "Wie entsteht eine imperiale Infrastuktur?," 58, 62–63.

74. For an overview, see Geo A. Schmidt, *Das Kolonialwirtschaftliche Komittee: Ein Rückblick auf seine Entstehung und seine Arbeiten aus Anlass des Gedenkjahres 50 jähriger deutscher Kolonialarbeit* (Berlin: 1934).

75. Otto Warburg, "Warum ist die Errichtung eines wissenschaftlich-technischen Laboratoriums in dem botanischen Garten zu Victoria erforderlich?," *Tropenpflanzer* 3, no. 7 (1899): 294.

76. Warburg, "Warum ist die Errichtung," 295.

77. On further influence of Dutch agronomy in Cameroon, see R. Thillard, "La Culture du Tabac de Sumatra au Cameroun," *Agronomie Coloniale* 40 (1921).

78. "Rapport présenté aux Chambres par le Ministre des Colonies," *Bulletin agricole du Congo belge* 1 (1910): 15–16.

79. L. Pynaert, "Le Jardin d'Eala," *Zooleo* 37, no. 3 (1957): 211.

80. Emile de Wildeman, *Mission Emile Laurent (1903–1904)* (Brussels: Vanbuggenhout, 1905), cxliv.

81. *Bulletin agricole du Congo belge* 2 (1911).

82. "Rapport présenté aux Chambres par le Ministre des Colonies" (1910): 11, 16, 320.

83. *Bulletin agricole du Congo belge et du Ruanda-Urundi* 1 (1910): 14, 16.

84. "Early History of Buitenzorg Botanic Gardens," *Bulletin of Miscellaneous Information (Royal Gardens, Kew)* 79 (1893): 173–75.

85. "Early History of Buitenzorg Botanic Gardens," 248, 253.

86. Auguste Chevalier, "Alerte aux plantations de Cacaoyers dans l'Ouest africain," *Revue internationale de botanique appliquée et d'agriculture tropicale* 26 (May–June 1946): 283–84.

87. The links between British West Africa and Buitenzorg need further research. These "hints" on the role of Buitenzorg are taken from C. J. J. van Hall, *Cocoa* (London: Macmillan, 1914), vi, 146, 382–92.

88. Auguste Chevalier, "Le Fonctionnement du Laboratoire d'Agronomie Coloniale," *Revue de Botanique Appliquée et d'Agriculture Tropicale* 1 (1923); Auguste Chevalier, "Historique de la Revue de Botanique Appliquée et d'Agriculture Tropicale," *Revue de Botanique Appliquée et d'Agriculture Tropicale* 23, no. 1 (1943): 1.

89. Christophe Bonneuil, "Auguste Chevalier, savant colonial: Entre science et empire, entre botanique et agronomie," in *Les sciences coloniales: figures et institutions*, edited by P. Petitjean (Paris: ORSTOM, 1996), 20.

90. Warburg, "Laboratoriums," 296. In 1902, Warburg studied different laboratories all over Europe and recommended copying the Dutch system. See Otto Warburg, "Über wissenschaftliche Institute für Kolonialwirtschaft," *Verhandlungen des deutschen Kolonialkongresses 1902* (Berlin: 1902), 202.

91. About William Howard Taft and the role he played in helping to establish the League of Nations, see Frank Gerrity, ed., *Taft Papers on League of Nations*, vol. 7, *Collected Works of William Howard Taft* (Athens: Ohio University Press, 2003).

92. Merrill, *Report on Investigations*, 7–10.

93. Warwick Anderson, *Colonial Pathologies: American Tropical Medicine, Race, and Hygiene in the Philippines* (Durham, NC: Duke University Press, 2006), 111.

94. He continued, however, to import seeds from Buitenzorg. Archives of the Arnold Arboretum of Harvard University: I B EDM, Section III Memoranda, Merrill to I. W. Bailey on shipment of Malayan woody specimens from Buitenzorg. See also William J. Robbins, *Elmer Drew Merrill, 1876–1956: A Biographical Memoir* (Washington, DC: National Academy of Sciences, 1958).

95. Stoler, *Capitalism and Confrontation*.

96. Maat, *Science Cultivating Practice*, 54; F. A. F. C. Went, "A Short History of General Botany in the Netherlands Indies," in *Science and Scientists in the Netherlands Indies*, edited by Pieter Honig and Frans Verdoorn (New York: Stechert, 1945).

97. BArch, N2303/1 Entwurf Tagebuch Stuhlmann Britisch-Indienreise/Bericht an Kolonialwirtschaftliches Komitee über die vom KWK finanzierte Reise nach Britisch Indien, Nr. 246.

Discovery and Patriarchy

Professionalization of Botany and the Distancing of Women and "Others"

CAREY MCCORMACK

The historiography surrounding Joseph Dalton Hooker's botanic explorations is extensive but not exhaustive. Nineteenth-century historians focused on his work with the Royal Society in London and the Royal Botanic Gardens at Kew, or they dealt with his travels in Northern India and the Himalayas as a heroic discoverer.[1] Recent scholarship complicates these early interpretations by connecting the work of sciences such as botany with the economic expansion of empire.[2] New research on indigenous knowledge and plant exchanges between indigenous people and British collectors explores the importance of indigenous engagement in networks of exchange, emphasizing the important role indigenous guides played as mediators of cross-cultural exchange.[3] This chapter diverges from previous historical scholarship by focusing on Joseph Hooker as a case study for the shift in botanical "discovery" in the nineteenth century from a large and diverse network of exchange to one centered and controlled by professional botanists, thus excluding formerly important participants. However, this study seeks to contribute to more recent scholarship on indigenous guides and assistants in order to challenge the image of the solitary traveler and instead emphasize the significant role of local people, common soldiers, and colonial wives.[4]

By the time Hooker traveled to the Himalayas, there was a diverse network of amateur and professional botanists around the world working toward classifying and naming plants within (and occasionally beyond) the British Empire. This network included British soldiers, colonial wives, Indian artists, and local indigenous botanists and plant collectors. While Hooker did not create this vast network of knowledge and plant exchange, he did shape it in an increasingly exclusionary way in his later years as director of the Royal Botanic Gardens, Kew. Hooker excluded certain groups of people, such as colonial wives and local merchants, based on categorizations of gender, race, and class. Hooker's later role as the director of Kew included the implementation of economic botany on a global scale. Harnessing plants for natural resource extraction in the late nineteenth century contributed to the ecological imperialism that is the central focus of this volume.

In the nineteenth century, universities in Glasgow, London, Cambridge, and Oxford began to offer degrees in botany, which led to a new class of professional botanists that replaced amateur plant enthusiasts who were vital to the networks of exchange. Individual collectors who previously claimed credit for the discovery of new species now had to send their collections to experts at Kew or Calcutta for classification and naming by these new professionals.[5] Experts excluded or replaced women, soldiers, and indigenous botanists based on these groups not having the "proper" education to claim discoveries. In this sense, the professionalization of botany, as evidenced by Joseph Hooker's letters sent while traveling in India, led to the network of botanical exchange centering in Kew around a traditional patriarchal figure who claimed the expertise to create and disseminate knowledge across the British Empire. This is not to say that all botanical knowledge formed around a patriarchal head, but that Joseph Hooker specifically organized the Indian network of exchange in this way to reposition India as the peripheral zone and harbor the knowledge created in the periphery at the imperial core.[6] Joseph Hooker attempted to organize and control a vast network of collectors that worked autonomously before his expeditions. Although Joseph Hooker is one man in a world full of botanists collecting during this time, his experience emphasizes the growing divide and exclusion of people based on gender, race, and class from scientific fields such as botany and the restructuring of global exchange around imperially controlled centers such as botanic gardens.

Joseph Hooker provides a useful case study on the shift toward exclusion for several reasons. The letters he sent during his travels in the Pacific, Antarctic, and Indian Ocean were widely published by his father, William Jackson Hooker, who was the director of the Royal Botanic Gardens in Kew before his son took office, with minor editing to flatter patrons and leading scientists of the time. The first section of this chapter discusses the differences in published representations of Joseph Hooker in comparison with his private descriptions based on the letters he sent. The analysis of Hooker's travels focuses on a comparison between his personal letters and his published narrative in order to problematize his image as a heroic and solitary plant collector.[7] Hooker's private letters reveal his opinions of the people working with him and within the network. In the published narrative, these texts are often excluded. The letters reveal his biases concerning amateurs and his work to organize the network of collectors in India based on social factors such as profession, education, and gender. Hooker also shifted his focus of discovery away from unique species and instead collected useful specimens that would further economic botany. This shift emphasizes the growing connection between science and the expansion of empire, but also represents the growing divide between people who ruled the empire and therefore could create knowledge useful to the empire, and who should be marginalized within the empire. In other words, the creation of knowledge through classification of plants transformed indigenous/local knowledge into colonial power by using local plants as commodities. Botanists' codification of local knowledge erased the origins and cultural significance of plant knowledge that came from local collectors and indigenous peoples. The shift from plant collecting as scientific curiosity to plant collecting for economic exploitation became Hooker's legacy and is the focus of the second section of this chapter. Furthermore, Joseph Hooker was one of the new professional botanists trained for the express purpose of collecting/discovering new species. His expertise allowed him to negate the work of others and to codify indigenous plant knowledge as Western scientific discovery. The third section of the chapter focuses on the local British network of collection in India and Hooker's instructions for these "hobbyists." Despite the growing exclusion of various groups of people from the act of collecting, Hooker continued to gather indigenous knowledge of plants and charged all collectors with the same task. It will later be argued that indigenous knowledge was, and continues to be, central to the development of botanical knowledge in South Asia despite the codification of such knowledge as Western scientific discovery. Finally, this study discusses Hooker's need for and use of local collectors as well as the various issues with collecting, including local resistance.

Joseph Dalton Hooker: Heroic Plant Hunter

Literary representation, in the form of articles and books in the nineteenth century, often presented authors in contrary terms to who the author was or how this person thought of themselves and others. For example, in 1849 William Taylor (the postmaster general of India during this time) created a series of watercolor paintings in Darjeeling specifically to record "the habits & manners of the English and the natives," in the Himalayas.[8]

One particular painting represents the work of Joseph Hooker collecting in Darjeeling. He is seated and dressed in a mixture of Himalayan and English garb while a local headman and two women are seated before him presenting plants in upraised hands. The painting represents local people in positions of reverence to Hooker who is situated in a position of authority. When describing the painting to his future wife, Frances Henslow, in his letters, Hooker describes himself as the specimen of the painting and that his clothes convey the exotic place he is in, but that the English clothes temper the exotic by emphasizing his authority in the painting. He believed that the painting was a romanticized version of his work in India but also that the topic of the painting, a day of collecting, was accurate.[9]

The painting clearly shows local people giving plants to Hooker and yet he describes the painting as his work for the day. In other words, Hooker believed that the plants were part of his collection even though he was given the plants by local people and indigenous guides. Hooker is surrounded by people, suggesting that plant collectors worked with assistants including help from local people knowledgeable of surrounding flora. However, in his private letters, Hooker fashioned himself as a professional botanist and negated the work of local collectors, leaving him as the sole authority in naming and classifying what others found.

Recent scholarship on the history of empire and science problematizes the heroic narrative of the professional male scientists and the utilitarian nature of science described in nineteenth-century travel narratives of plant collectors.[10] Early works on Joseph Hooker exemplify the glorification of professional botanists and their contributions to science.[11] By the time Leonard Huxley wrote his sprawling two-volume biography on Joseph Hooker, authors such as Huxley described scientists as selfless, intelligent men who civilized the world by spreading scientific knowledge within the British Empire. When describing the time period of the 1860s, Huxley argued that "it was the century of the New Renaissance. The revolution in thought was paralleled by a revolution in the means of civilized life. . . . Power over nature transformed the way of life."[12] As the director of Kew, like his father before him, Hooker consolidated the vast network of botanists and directed their collecting around economic botany. This included turning botanic gardens around the world into plant research stations, something Joseph Banks began in the eighteenth century. This consolidation around one main (patriarchal) figure mirrors the consolidation of power in the empire and the growing representation of the British Empire as a civilizing, paternal force.

Hooker, and many lesser known botanists around the world, developed extensive correspondence with famous botanists such as Alexander von Humboldt as a way to legitimize their findings.[13] William Hooker published his son's letters in his *London Journal of Botany* in order to familiarize the larger community of gentlemen scientists and plant enthusiasts with Joseph Hooker's travels. One such letter, between Joseph Hooker and Alexander von Humboldt, emphasized the shrinking field of botany by excluding amateur collectors and placing Hooker and other professional botanists within a privileged class capable of creating knowledge about the natural world. Humboldt described Joseph Hooker as a "noble traveler" collecting new species and information to include in the expanding field of professional botany.[14] Publications in journals such as this provided readers with an introduction to new botanists and their findings. Both Joseph Hooker and Brian Houghton Hodgson, working in Darjeeling before Hooker's arrival, confirmed many of Humboldt's postulations regarding the Himalayas, suggesting that Humboldt was considered by many naturalists to be an expert on Central Asia despite his never having traveled in the region.[15] Famous botanists such as Humboldt influenced how scientists like Hooker observed and created knowledge in the Himalayas. Similarly, Hooker and Hodgson collected based on previous excursions into the region conducted by Nathaniel Wallich in Nepal from 1820 to 1821. Wallich also depended on botanists before him and advanced his herbarium through the inheritance of William Roxburgh and Francis Buchanan-Hamilton's plant collections in Calcutta.[16] Hooker's letters constantly reference the genera and species of plants "discovered" by Wallich, even though Wallich's own letters reveal his dependence on local collectors and their knowledge.[17] Combining collections provided extensive knowledge to whichever institution gained the collection, but credit for discoveries rarely named other people besides the botanist in possession of the specimen. Hooker combined his collection with Dr. Thomson, who worked at a Tibet mission, combining 4,000 plants that were transported together to Kew in London.[18] While it is clear that combining collections in this way was economical for transportation, deciding who would take credit for the discovery depended on who claimed professional status.

Joseph Hooker was shocked to find that he was famous in India due to publications of his work in the Pacific and Antarctic written in letters by another officer of the ship named Sir John Ross in the *London Journal of Botany*.[19] Publications informed hobbyists and professionals alike across the expanse of the British Empire. Hooker found it amusing that botanical enthusiasts in India were familiar with his work to an extent that people in England were not. This not only reveals the transfer of knowledge around the globe but also who had access to this knowledge.[20] The people who were familiar with Hooker's work had access to journals such as this because they had the money to purchase these works and also the literacy and education to interpret the information. Many historians have studied the rise in print culture, literacy, and education so I will not provide a discussion of this topic here. However, the journals highlight how botanical interests and collecting became exclusionary during this time period, providing access to new botanical knowledge for middle-class, educated men with the means to purchase and engage in this network of knowledge and thereby excluding women, poor British men, and indigenous scientists.[21]

While botanic journals serve as one example of the growing exclusivity of botany, the naming and categorization of plants provides a gendered perspective of this shift. Since the eighteenth century and the creation of networks of plant exchange by Joseph Banks (through the Royal Society and Kew) and the East India Company, people from around the world collected plants and sent them to the Kew gardens as curiosities or economic crops.[22] It was at this center of exchange that collectors classified plants based on a patriarchal system of patrimony. Active participants in botanical collection sent their finds to centers of botanic knowledge so professional (male) botanists could properly categorize and name each plant.[23] The network of exchange and diffusion of knowledge and plants became centered on stratified and gendered claims to scientific knowledge. This codification of knowledge linked the colonization of local knowledge with patrimonial naming traditions.[24]

The process of naming developed around honoring botanists, patrons (especially the aristocracy), and occasionally plant collectors. For example, the entire genus of *Helwingia* was named after Decaisre, the "founder of the Order," according to Hooker.[25] In conjunction with the act of naming as an honorific, botanists who published their work dedicated books to those who assisted plant collectors and travelers such as Hooker. In an account of Joseph Hooker's mission in India, William Hooker introduces the private letters from his son by acknowledging the contribution of Countess Dalhousie and her herbarium and plant collection in India.[26] This acknowledgment contrasts with the letter sent by Joseph Hooker to Frances Henslow, in which he describes Lord and Lady Dalhousie having "a most unfortunate indifference to science in any shape," and instead the couple talked mostly of aristocratic gossip.[27] In an earlier letter, Hooker describes "Lady D" as vapid and practically illiterate, as was her naturalist brother, Lord Arthur Hay.[28] Perhaps Hooker represented Lord Hay in an unsavory light because Hay was a private collector with an interest in natural history but did not have a professional education in botany.[29] This reveals the distancing of botanists such as Hooker from individual enthusiasts such as Hay, while the dedication made by William Hooker emphasizes the importance of aristocratic patronage during the mid-nineteenth century. Whatever Joseph Hooker's opinion of Lord and Lady Dalhousie, they traveled with Hooker and provided him with introductions and protection while in India.

As Hooker's letters reveal, botanists juggled between gathering new and unique specimens and collecting economic crops. However, Joseph Hooker's legacy, according to Huxley's biography, was not his discovery of new species of rhododendrons, orchids, and azaleas. Instead, Hooker became famous for his work on economic crops such as cinchona and rubber trees, as well as edible crops such as rice, maize, and barley. This is significant because the focus on economic botany connects plant collecting to the expansion of empire. Hooker's goal was not merely to "discover" new plants, but to collect local crops and information about the indigenous methods of agriculture. This knowledge could then be used to further the economic viability of British India and fuel expansion. For example, Hooker describes the cultivation of rice at high elevation, detailing the lack of irrigation and increased soil erosion in South Asia.[30] In another letter to Frances Henslow, Hooker described in detail the various crops grown at high elevation such as buckwheat, barley, rice, mangos, turnips, and radishes.[31] These details and descriptions of food crops are more than curiosities related to a loved one. Instead, information regarding the erosion, elevation, and seasonal crops were later used to produce tea plantations in Darjeeling.[32] Hooker's travels in the Himalayas can be construed as imperial reconnaissance not only for his part in gathering local agricultural knowledge but also for future markets. He informed his father that one of his goals in traveling to Tibet through Bhutan was to provide a new market for goods at a quarter of the price of Chinese merchants in the area.[33]

As a proponent of economic botany, Hooker encouraged each botanic garden in British India to develop research centers and experiment with cash and drug crops such as cinchona and tea.[34] Hooker lamented to his father that local people used the Calcutta gardens as a social place rather than "a place for profit & instruction."[35] As director of Kew later in life, Hooker successfully transformed British botanic gardens around the world into experiment stations and centers for exchange. Furthermore, his travels connected botanist interests with the manufacturing of opium in India. Used as a cash crop for many years, Hooker asked the owner of an opium factory to provide "specimens of all the articles & implementations employed" in making opium.[36] Again, this is beyond scientific curiosity and instead entails active participation in the economic expansion of empire given the opium wars fought and won by the British, a few short years previous to Hooker's travels.

Any plant could be used as a form of payment during this time period, but most collectors specifically looked for fruits, vegetables, spices, and drugs.[37] While in Bhagalpur, Hooker asked Ross, the curator of the garden there, to collect "Fruits & Economic products for me," which Hooker would transport back to England on his return journey.[38] Collectors transported economic plants around the world since the time of the Columbian exchange, but the purposeful collection of fruits and vegetables furthered colonial expansion by providing food for settlers and colonists. For example, Hooker describes the abundant supply of plantains, bananas, bread fruit, pineapples, and betel palms in Ceylon, none of which are native to the island.[39] Botanical ships and settlers transferred those plants during the eighteenth and nineteenth centuries in order to provide food for the growing populations within the colonies. However, it is possible that several of the local varieties were transferred to the islands by pre-European exchange between the Sinhalese and Tamils.

Organizing the Network

As discussed above, there were many independent British collectors in India, ranging from East India Company servants, to soldiers, surgeons, and wives.[40] All of these amateur botanists and collectors sent plants, drawings, seeds, and information on propagation and care to professional botanists at Kew and local botanic gardens such as in Calcutta. Hooker describes his encounters with many of these unofficial collectors and often sent their local knowledge to his father. In one such letter, Hooker discussed the information he gathered concerning the proper soil needed for cotton cultivation in Mysore, which he obtained from General Cubbon's surgeon.[41] Hooker employed the help of collectors such as Major Napleton in Monghyr, who collected in the nearby jungles during his free time.[42] All manner of people collected and experimented with local species either as a hobby or to learn the economic benefits of species. Whether low-ranking soldiers or officers, Hooker employed all plant collectors that he met to gather independently and send their collections back to England or to local botanic gardens in Madras and Calcutta for scientific classification.[43]

The network of amateur and professional botanists in India reflects the diversity of interest in plants. Hooker often mentioned the lack of botanical (professional) knowledge many collectors had and chastised men who held prestigious positions within botany who did not know local uses or names. For example, Hooker criticized the director of the botanic garden in Calcutta,

John McLelland, by stating, "he [McLelland] does not know the name of any of the garden plants & always takes a Molli (or black gardener) to tell me the names."[44] This lack of knowledge was excusable among amateur hobbyists but not the director of the gardens. It is also interesting that Hooker did not state the name of the Indian gardener even though he was clearly the expert in this situation. This reference to local gardeners prevails in nineteenth-century letters, but few English naturalists or botanists name these people and focus instead on the information provided by indigenous people. While most local guides, gardeners, and collectors provided extensive knowledge and specimens of local flora, they are only ever represented as servants or hired help.[45] Their presence in the historical record challenges this representation, especially since botanists like Joseph Hooker purposefully acquired their help in providing local botanical knowledge. This active exclusion was most likely due to racial and gendered classifications discussed above.

Even more disconcerting than the lack of knowledge McLelland professed was the conflict of interests and goals among professional botanists in India and the larger empire as botany developed into a professional field of scientific inquiry. McLelland provides one example of how government agents, professional botanists, and independent collectors clashed with one another. He ignored the previous director's instructions on maintaining the garden and often built additions in opposition to the East India Company's needs for an experimentation station. Nathaniel Wallich, for instance, also collected according to his own interest and in turn shunned the directions of the East India Company, making the task of procuring funds from the company for plant collection increasingly difficult for later botanists. Despite the vast network of botanists in communication with one another, disagreements and divergence from the stated path of economic botany forced plant collectors to work independently, oftentimes lacking funds to continue. State-sponsored collecting such as Joseph Hooker's work in Northern India was much more successful than independent collectors for this reason.

Although the field of botany and those involved in categorizing and naming plants declined during this time period, the collection of indigenous plant names and uses continued. Hooker searched specifically for indigenous experts during his travels. While a small group of British men obtained authority over who "discovered" plants, professional botanists such as Joseph Hooker charged their local networks of collectors with obtaining as much indigenous knowledge as possible.[46] In a letter to his father, Hooker referred to Griffith's work post mortem, "the first thing published is his travels & journals: —It is to be regretted there are no notes explanatory of old Indian names & terms, nor chart or map. (This is McLelland's great fault & a most serious want: an Indian (as McL. is) only can supply this deficiency & any well informed Indian could.)"[47] Many botanists working in botanic gardens in India died before publishing their work, leaving their predecessors to interpret their work and organize the collection for publication. Hooker found McLelland's work lacking relevant local information, which proves that indigenous knowledge continued to be useful and sought after by professional collectors but also reveals that collectors did not so much discover new plants as gather plants well known to local people and codify new specimens within the Linnaean taxonomy that tended to erase local names, uses, and their cultural significance.[48] Hooker brought seeds and drawings to local markets during his travels in India for identification by locals. He also collected grains and drugs in the markets along with local cultivation techniques and the Hindu, Persian, and English common names of these plants.[49] His purpose in collecting local knowledge was to convert indigenous terms to their "proper order"—in other words, codifying and thus claiming indigenous knowledge as European science. Even when he respected the people who provided knowledge, his tone conveys his belief in their ignorance of scientific botany, which immediately discredited their equal participation in the act of discovery. While traveling in Tibet, Hooker conversed with local monks for information on the Yaron River, stating, "I am endeavoring to obtain from the Lamas &c, who come across the snow, all the information they can give about the crops on the route to China, through countries of which we know nothing."[50] But even these "well-educated" monks were not legitimized in their knowledge of the region until Hooker spoke to many "informants" and was satisfied that they all provided the same information.[51]

Local Collectors

Hooker's emphasis on obtaining indigenous knowledge involved hiring local collectors skilled at botany. They were tasked with gathering specimens, drying them, and drawing/painting representations of the "discoveries."[52] The category of local collector refers to indigenous people of India and the Himalayas hired for the purpose of collecting plants, seeds, and specimen, and information regarding their uses. This does not mean these collectors were from the areas under examination, but that Hooker obtained collectors when he needed them and he expected those collectors to travel with him until his journey ended. It should also be noted that he did not choose collectors based on the convenience of their location, but for their knowledge of local flora and possibly for their skills in preserving specimen. In several letters to his family, Hooker expressed his frustration at not finding skilled collectors and that the collectors he did hire were referred to him by other British collectors in India.[53] Hooker's intention was to hire several plant collectors with experience and who would accept 20 rupees a month in pay.[54] Because of this budget, Hooker intended to have no more than five local collectors travel with him through Darjeeling, Tibet, and Sikkim.

Although some collectors were not indigenous to the mountainous regions in the north, their skills in identifying and collecting plants were essential to Hooker's success. Like many European travelers in South and Southeast Asia, Hooker planned to travel by palanquin, which left the work of collecting on the journey to local guides and collectors. Indeed, his plan was to travel while "2 plant collectors, will be sent on before."[55] Hooker's plans quickly unraveled as the number of "discoveries" increased and therefore his need for local plant collectors.

Hooker's retinue grew exponentially once he reached the Himalayas. He hired local experts as guides, servants, carriers, and plant collectors. By August 1848, Hooker had doubled his retinue of collectors to almost twenty. Each day, collectors set out in different directions, sometimes "across the snow," to collect "roots of Rhododendrons & whatever else may be in flower" and press each specimen in brown paper for preservation.[56] It is clear from these letters that indigenous plant collectors worked autonomously and used personal initiative in gathering specimens. While not everything collected was a new discovery, there were enough to make these collectors botanical assistants at the least, and botanic specialists in several cases. And yet, Hooker, like many professional botanists during the mid-nineteenth century, rarely names these collectors or credits them with the discovery of new species.[57] Their work is silenced in most biographies discussing Joseph Hooker's travels and yet the number of twenty collectors seems noteworthy considering the vast amount of species "collected" (some 4,000 in one year according to Hooker).

While Hooker traveled in the Himalayas, he asked local people in Madras and Calcutta to collect independently so that he could retrieve the specimens on his return journey. In this way, Hooker actively created a network of collection in India that provided him with plants that he then brought back with him to England for classification, naming, and experimentation. In other words, he organized an existing network of collectors in India by charging each member in the network to send on their specimens to Kew Gardens for "expert" analysis. Once at Kew, professional botanists codified the plant by providing it with Latin nomenclature and added the plant to geographic lists. The person who sent the plant had no claim once the process was complete.

This exclusion is evidenced throughout Hooker's private correspondence. While Hooker mentions some of the names of local British and indigenous collectors, none of these participants received credit for the "discovery" of new species. This mirrors the shift from botany as a common interest and hobby to the field ruled by the professional. It was no longer the prerogative of the finder to name their discoveries, but instead all new species had to be sent to the appropriate stations and all knowledge sent to the appropriate specialist, who would then create empirical (and perhaps imperial) knowledge concerning the plant. Botany, like many other fields of science and technology in the nineteenth century, became a white male–dominated field that was used as a tool of dominance over colonial subjects and women.[58] These new professionals exerted dominance by excluding women and indigenous plant collectors from the professional field of botany and yet botanists processed the discoveries made by these excluded groups through a knowledge-producing colonial machine (botanic gardens).

Exclusion from science included patronizing indigenous collectors and servants in order to position locals as scientifically backward. Joseph Hooker praises and complains about his servants and collectors, depending on the ethnicity and tasks of each. For example, Hooker trusted one collector he named as Clamanze to carry his collection to Calcutta.[59] This same collector Hooker rewards for his dedicated work, stating, "my serv[an]t is so good that I bought him a poney [*sic*] to save his long walks & carry the collecting papers."[60] Hooker does not discuss the heritage of this servant so it is possible that Clamanze is Goan, Portuguese, or Spanish-Philippine, but it is clear that Hooker did not consider his work to be anything more than a servant's duties, albeit a dedicated one. However, if collectors were sent out on autonomous missions to gather roots, seeds, and cuttings then it can be argued that Clamanze found new species unknown to his Western employer. The lack of archival sources detailing the personal information of collectors or any temporary employees of gardens and plant collectors make conclusions about the nature of their role in the expansion of botanic knowledge conjecture. What is clear from the record, and specifically from Joseph Hooker's letter, is that local plant collectors worked more as research assistants than as servants, despite Hooker's representation of Clamanze as a servant.

Praise for the collectors is rare in the letters as Hooker traveled further into the Himalayas. He complained to Frances Henslow that the collectors he sent "beyond the snow" in Nepal brought back poor samples and several plants that had already been collected. He worried that collectors did not go as far as he instructed which is why they produced duplicates.[61] Since Hooker was not venturing with these collectors, he was unable to manage their work and his first instinct was to distrust them. Later in the letter he states, "my collectors, miserable Bengalees [*sic*] who took me to Tonglo & were utterly useless, either at gathering plants or anything else, have sickened of fever & ague though I paid them all possible attention."[62] This letter influenced Hooker's later suggestions to the director of the botanic garden in Calcutta, Dr. Anderson, to gather seeds personally and forego the use of indigenous plant collectors.[63] The harsh opinions described above clashed with his earlier frustration at not finding enough plant collectors to take on his journey north. It is possible that plant collectors from Southern India and Bengal would not be familiar with the flora of the Himalayas. However, Hooker's frustration is less with collectors whom he hired from local societies in the mountainous region than those he brought from Southern India.

Local Resistance

It is also possible that collectors who ignored the directions of Hooker or brought back duplicate specimens were purposely hindering Hooker's work. Hooker did not fail to recognize local resistance to his efforts and often spoke of the difficulties he encountered with weather, local governments, and communication. Early in his travels, Hooker complained that he had done much "botanizing" in Aden on his journey to India, but the saltwater and cramped quarters of the ship destroyed the specimens he gathered. Transportation of collections proved a significant obstacle from the first sea voyages sponsored by Joseph Banks to the passenger and cargo ships carrying vast collections across the British Empire in the nineteenth century.[64]

Botanists such as Hooker, traveling from London to India, often stopped in many places before reaching the final destination, which gave botanists the opportunity to collect in diverse regions as well as drop off parts of their collection to other botanists and botanic gardens. For example, Hooker stopped in Egypt, Aden, and Ceylon on his way to India and visited Borneo and other Southeast Asian islands on his return journey. In each of these places, Hooker used the opportunity to collect plants and local knowledge, as well as draw specimens when collecting would be useless. However, the colonies ruled by European powers in places categorized as tropical, like Ceylon, did not offer botanists such as Hooker a permanent home or station from which to work. Scientific societies and organizations in the mid-nineteenth century assumed that botanists would work temporarily in tropical zones but were unfit to live there.[65] Tropical areas were meant to be places of discovery, in which white male botanists could make a name for themselves and return to England for a comfortable position in Glasgow or London. Hooker exemplified this attitude when describing Ceylon: "these are nice places to see, but I am not inclined to live in; as the pale yellow & all but sickly faces of the English children plainly tell."[66] Race played a key role in perceptions of the colonies and what climates were suitable for British colonizers. For example, Hooker complained while traveling in the Himalayas, "the society is wretched; & I only care to know a few quiet people who can & do aid me, sending plants."[67] Since British colonization had expanded only into Darjeeling during this time period, it is clear that Hooker is describing the "society" of tribes he encountered on his travels. He did not wish to make indigenous people part of his scientific network of exchange. In this sense, his communication with native "informants" was precisely that: a means to extract information from indigenous sources that could then be transferred into imperial knowledge and British masculine discovery.

At least one indigenous group suspected Hooker's goal was to extract information for imperial gains, because the Rajah of Sikkim refused Hooker entrance to his country and when Hooker ignored the rajah and entered anyway, Hooker was imprisoned for a short period of time for trespassing.[68] Before this dramatic confrontation, the rajah sent Meepo, an indigenous guide, to Hooker as a means to control where the British botanist traveled but also to spy on Hooker and discover his (and Britain's) intentions in the region.[69] He was insulted that the rajah would send a man to deliberately lead his party astray and spy on their work. Beyond Meepo and the servants sent with him, Hooker described the people as having a "hostile disposition," to the botanic retinue although he does not distinguish which people he is describing.[70]

Just as the categorization of tropic and temperate climates determined which races could flourish in certain climates, according to nineteenth-century medical and scientific theory, race also influenced how botanists interacted with indigenous villagers and tribes in Northern India. Hooker used the terms "very black," "uncouth," and "a very turbulent race of savages" when describing the people in "Dingcham province, Damtsen."[71] Characterizing the village as savage and black provided a binary contrast to Hooker's intended goal of advancing scientific knowledge of the region, something that would in turn provide advancement and "civilization" to spread with the expansion of empire.

Racial categorizations aided Hooker in assessing which tribes would help the botanist and which would not. He described the Lepcha and Limbo tribes based on their physical features and political loyalties. According to him, the Limbo tribe was allied with the rajah of Sikkim, which inevitably positioned them as enemies. Hooker employed many Lepcha people as collectors, servants, and bearers and even enjoyed their company as traveling companions. The Limbo received no such employment because in Hooker's opinion, they could not be trusted.[72]

However, Hooker and his collecting retinue were not the only people who saw a racialized and exotic "Other." While passing through a Lepcha village, the local inhabitants came out of their homes to watch Hooker and his retinue travel by. Hooker and other Europeans in his group gave the children silver coins and these were later made into necklaces and other jewelry. The collecting expedition was a strange spectacle walking past a village, which is something historians studying race and empire tend to ignore. Being prevented from crossing the river and the handing out of silver coins are part of a complex encounter between tribes in the Himalayas and European and Indian collectors. Communication barriers during such encounters influenced the experience that later impacted how Western scientists such as Hooker represented indigenous peoples. Helpful "natives" appear curious, while territorial tribes are represented as "savages" and "black." While there is no way to know how the Lepcha village interpreted the botanic expedition, it is clear by their distant watching that Hooker's party offered an exotic spectacle not seen by them before.[73]

Conclusion

Joseph Hooker provides a typical case study of a significant shift in the network of exchange from a diverse network of people engaged in botanical "discovery" to a white, male-dominated profession. While professional botanists such as Joseph Hooker relied on indigenous knowledge about cultivation, soil erosion, adaptation, edibility, and medicinal uses of plants collected in British holdings, local collectors who performed the majority of the work increasingly became silent partners in "discovery." Botany and the expansion of empire were intimately connected during the mid-nineteenth century and the hardening of colonial categories is evidenced by this shift toward botany as an exclusionary science. Indigenous knowledge and plant specimens were transformed into Western scientific excavation that contributed to the "process of global and ecological homogenization" as Brett

Bennett suggests in this volume. The diversity of the networks and the rate of exchange increased during this time as transportation by land and sea improved. However, exchange became centered at points of expertise encompassed by botanic gardens and experimentation stations, creating and supporting new imperial cores in botanic gardens in Calcutta and London. Collectors, hobbyists, and amateurs were instructed by professional botanists to send their findings to the experts to be properly categorized and named.

This study focused on the changing relationship between professional botanists and the networks of plant exchange in India during the nineteenth century. Freedom of exchange continued and professional botanists such as Joseph Hooker encouraged this. But the movement of plants and the production of plant knowledge changed during this time period by centering the formation of knowledge, and distribution, around a patriarchal head at Kew. Indigenous guides and collectors continued to be imperative in botanic exchange, yet their roles became increasingly deemphasized by professional botanists. Hooker provides a poignant example of the key role gender played in creating scientific knowledge. By the middle of the nineteenth century, men (particularly white European men) positioned themselves as the only true creators and purveyors of knowledge, thus ignoring the cross-cultural exchange of indigenous plant knowledge that was central to understanding nature. By silencing indigenous knowledge of plants, ecological sustainability, resistance, and local lifeways altered according to European needs to export natural resources from colonial holdings.

Notes

1. Some of these historians are addressed later in this chapter but the trend of presenting plant collectors as heroes continued well into the twentieth century. See Kenneth Lemmon, *The Golden Age of Plant Hunters* (London: Phoenix House Publications, 1968) and Ray Desmond, *Plant Hunting for Kew*, edited by Nigel Hepper (London: HMSO, 1989).

2. Londa L. Schiebinger, *Plants and Empire: Colonial Bioprospecting in the Atlantic World* (Cambridge, MA: Harvard University Press, 2004); Richard Drayton, *Nature's Government: Science, Imperial Britain, and the "Improvement" of the World* (New Haven, CT: Yale University Press, 2000); Lucile H. Brockway, *Science and Colonial Expansion: The Role of the Royal Botanic Gardens* (New York: Academic Press, 1979).

3. Shino Konishi, Maria Nugent, and Tiffany Shellam, eds., *Indigenous Intermediaries: New Perspectives on Exploration Archives* (Acton, ACT: Australian National University Press, 2015). This book focuses on Australian exchanges but the critique

and methodologies of working with British archives to obtain an indigenous voice emphasizes the growing "reinterpretations of exploration history" as stated in the introduction (p. 1).

4. For an analysis that concentrates on developments in the metropolis, see Ruth Barton, "'Men of Science': Language, Identity and Professionalization in the Mid-Victorian Scientific Community," *History of Science* 41 (2003): 83, 90.

5. Barton, "'Men of Science'," 74–75. Barton complicates the term "professionalization" and the binary categories of professional and amateur. However, this study focuses on one field of science and tracks the development of botany, which diverges from Barton's argument.

6. As many historians in this volume and the broader field of imperial history argue, many centers existed in the globalized network of exchange of plants, ideas, and animals. This study utilized Tony Ballantyne's *Orientalism and Race: Aryanism in the British Empire* as a theoretical framework, suggesting that there were several imperial cores (not just one in Europe), Calcutta being one very important center. This and other works utilize network theory in order to de-center Europe within a larger global framework. Tony Ballantyne, *Orientalism and Race: Aryanism and the British Empire* (New York: Palgrave Macmillan, 2002).

7. A prevailing theme in the historiography of Joseph Hooker is the representation of botanists as utilitarian scientists/heroes working to legitimize their fields through solitary collecting. See Jim Endersby, *Imperial Nature: Joseph Hooker and the Practices of Victorian Science* (Chicago: University of Chicago Press, 2008), introduction, for a complex discussion of scientific curiosity in contention with science as a profession/vocation.

8. Joseph Dalton Hooker to Frances Henslow, 25 April 1849, Hooker Collections, Royal Botanic Gardens Kew (RBK), London, UK; the painting can be found in Ray Desmond, *Sir Joseph Dalton Hooker: Traveler and Plant Collector* (London: Antique Collectors Club Dist., 2007).

9. Joseph Dalton Hooker to Frances Henslow, 25 April 1849, RBK.

10. Brett M. Bennett and Joseph H. Hodge, eds., *Science and Empire: Knowledge and Networks of Science across the British Empire, 1800–1970* (London: Palgrave Macmillan, 2011); Londa Schiebinger and Claudia Swan, eds., *Colonial Botany: Science, Commerce, and Politics in the Early Modern World* (Philadelphia: University of Pennsylvania Press, 2005).

11. Leonard Huxley, *Life and Letters of Sir Joseph Dalton Hooker*, vol. 1 (London: John Murray, 1918), viii.

12. Huxley, *Life and Letters*, 2:1.

13. Much has been written about Alexander von Humboldt as one of the leading plant collectors and botanists in Brazil. Recent scholarship argues that his work aided imperial expansion by providing reconnaissance of areas not colonized by Europeans. See Mary Louise Pratt, *Imperial Eyes: Travel Writing and Transculturation* (New York: Routledge, 1992).

14. Alexander von Humboldt to Joseph Dalton Hooker, 11 December 1850, in *Hooker's Journal of Botany and Kew Garden Miscellany*, vol. 3, ed. William Hooker (London: Reeve, Benham, and Reeve, 1851), 21.

15. Joseph Dalton Hooker to Alexander von Humboldt, 23 September 1850, in *Hooker's Journal of Botany and Kew Garden Miscellany*, 22.

16. David Arnold, *The Tropics and the Traveling Gaze: India, Landscape, and Science, 1800–1856* (Seattle: University of Washington Press, 2006), 152.

17. Joseph Dalton Hooker to William Hooker, in *Hooker's Journal of Botany and Kew Garden Miscellany*, vol. 2, 13–14.

18. Joseph Dalton Hooker to Alexander von Humboldt, 23 September 1850, in *Hooker's Journal of Botany and Kew Garden Miscellany*, vol. 3, 31.

19. Hooker discusses his shock at being famous in Joseph Dalton Hooker to William Hooker, 7 April 1848, Hooker Collections, Royal Botanic Gardens Kew, London, http://www.kew.org/science-conservation/collections/joseph-hooker/explore (18 February 2017).

20. Women were not banned from participation in the study and creation of scientific knowledge; however, they were delegated to the roles of teachers and commentators while men actively participated in scientific inquiry. This explains the dominant role of men as explorers and botanists, while contemporary women participated in horticulture and botany as hobbies. Marina Benjamin, "Elbow Room: Women Writers on Science, 1790–1840," in *Science and Sensibility: Gender and Scientific Enquiry*, ed. Marina Benjamin (Oxford: Basil Blackwell, 1991).

21. It should be noted that these excluded groups did participate in scientific print culture but most botanical publications targeted men specifically. Several horticulture journals and travel books targeted women audiences because these publications tended to have less scientific jargon and more descriptions of exotic landscapes. As for local papers, there were many in India published in local languages that included scientific discovery; but again, these publications catered to local people who most likely did not speak English fluently enough to read scientific journals in English.

22. Endersby, *Imperial Nature*, 62.

23. Women living in South Asia often engaged in gardening, plant collecting, and plant experimentation. Gardening, as a hobby, connected to stereotypes of genteel society and femininity. Tracey Rizzo and Steven Gerontakis argue that "insect and plant specimen collecting was a legitimate pursuit for a Victorian woman of means. It signaled her engagement, not with peoples and cultures, but with the delicate side of nature. It was one of the major ways women both benefited from and facilitated imperialism." Tracey Rizzo and Steven Gerontakis, *Intimate Empires: Body, Race, and Gender in the Modern World* (Oxford: Oxford University Press, 2017), 95.

24. In the Linnaean system of classification, plant collectors chose the name of the species and subspecies of plants, which provided them opportunities to name species after people. For example, in 1818 Joseph Arnold collected in Sumatra with Stamford Raffles. Together they discovered a giant flower, so Arnold named the genus Rafflesia in honor of Raffles. See Philip Short, in *Pursuit of Plants: Experiences of Nineteenth and Twentieth Century Plant Collectors* (Portland: Timber Press, 2003), 47–50.

25. Joseph Hooker to William Hooker, in *Hooker's Journal of Botany and Kew Garden Miscellany*, 2:12.

26. William Hooker, *The London Journal of Botany*, vol. 7 (London: H. Bailliere, 1848), 6.

27. Joseph Hooker to Frances Henslow, 15 January 1848, RBK.

28. Joseph Hooker to Elizabeth Hooker, 10 January 1848, RBK.

29. Endersby, *Imperial Nature*, 2.

30. Joseph Hooker to William Hooker, in *Hooker's Journal of Botany and Kew Garden Miscellany*, vol. 2, 20.

31. Joseph Hooker to Frances Henslow, 26 May 1848.

32. Huxley, *Life and Letters*, 17.

33. Joseph Hooker to William Hooker, 16 March 1848.

34. Huxley, *Life and Letters*, 2.

35. Joseph Hooker to William Hooker, 16 March 1848.

36. Joseph Hooker to William Hooker, 7 April 1848.

37. Endersby, *Imperial Nature*, 84–85.

38. Joseph Hooker to William Hooker, 7 April 1848, RBK.

39. Joseph Hooker to Frances Henslow, 15 January 1848, RBK.

40. David Mackay, "Agents of Empire: The Banksian Collectors and Evaluation of New Lands," in *Visions of Empire: Voyages, Botany, and Representations of Nature*, edited by David Philip Miller and Peter Hanns Reill (Cambridge: Cambridge University Press, 1996), 40.

41. Joseph Hooker to Frances Henslow, 10 January 1848.

42. Joseph Hooker to William Hooker, 7 April 1848.

43. Matthew Laubacher, "Cultures of Collection in Late Nineteenth Century American Natural History" (PhD diss., Arizona State University, 2011), 11.

44. Joseph Hooker to William Hooker, 20 January 1848.

45. Konishi, Nugent, and Shellam, *Indigenous Intermediaries*, 5.

46. Endersby, *Imperial Nature*, 89.

47. Joseph Hooker to William Hooker, 20 January 1848, RBK.

48. In a letter to Humboldt, Hooker assured his reader that he actively collected, "oral information . . . about its native and cultivated vegetation," in order to determine the geography of one specific Tibetan mountain region. Joseph Dalton Hooker to Alexander von Humboldt, 23 September 1850, in *Hooker's Journal of Botany and Kew Garden Miscellany*, 26.

49. Joseph Hooker to William Hooker, 16 March 1848, RBK. Although Hooker constantly referred to the local names and uses of plants, he categorized and named plants based on the Linnaean system, which favored Latin terms over indigenous names.

50. Joseph Hooker to Elizabeth Henslow, 26 May 1848, RBK.

51. Joseph Dalton Hooker to Alexander von Humboldt, 23 September 1850, in *Hooker's Journal of Botany and Kew Garden Miscellany*, vol. 3, 27.

52. Joseph Hooker to William Hooker, 16 March 1848, RBK.

53. Joseph Hooker to William Hooker, 16 March 1848, RBK.

54. Joseph Hooker to Elizabeth Hooker, 10 January 1848, RBK.

55. Joseph Hooker to William Hooker, 20 January 1848, RBK.

56. Joseph Hooker to William Hooker, 9 August 1848, RBK.

57. Arnold, *Tropics and the Traveling Gaze*, 183.

58. Michael Adas, *Machines as the Measure of Men: Science, Technology, and Ideologies of Western Dominance* (London: Cornell University Press, 1989), 14.

59. Joseph Hooker to Lady Hooker, 26 April 1849, RBK.

60. Joseph Hooker to William Hooker, 16 March 1848, RBK.

61. Joseph Hooker to Frances Henslow, 26 May 1848, RBK.

62. Joseph Hooker to Frances Henslow, 26 May 1848, RBK.

63. Huxley, *Life and Letters*, 8.

64. Mackay, "Agents of Empire," 43.

65. For a greater examination of this topic, see Mark Harrison, *Climates and Constitutions: Health, Race, Environment and British Imperialism in India, 1600–1850* (New Delhi: Oxford University Press, 1999).

66. Joseph Hooker to Frances Henslow, 15 January 1848, RBK.

67. Joseph Hooker to Frances Henslow, 26 May 1848, RBK.

68. Hooker's imprisonment has been discussed at length by other historians so no further details will be provided here. For a full account of the controversy, see Arnold, *Tropics and the Traveling Gaze*, 216–20.

69. Joseph Hooker to Lady Hooker, 24 May 1849, RBK.

70. Joseph Hooker to William Hooker, 14 October 1849, RBK.

71. This group of people attempted to stop Hooker and his retinue from crossing the river, which may have influenced his representation of them. Joseph Dalton Hooker to Baron von Humboldt, 23 September 1850, in *Hooker's Journal of Botany and Kew Garden Miscellany*, vol. 3, 25–26.

72. Joseph Hooker to William Hooker, in *Hooker's Journal of Botany and Kew Garden Miscellany*, vol. 2, 19–20.

73. Ingjerd Hoem, "The Scientific Endeavor and the Natives," in *Visions of Empire*, 320–22.

PART III | Animal Agency

Animal Skinners

A Transcolonial Network and the Formation of West African Zoology

STEPHANIE ZEHNLE

When, in the 1880s, the young Swiss biologist Johann Büttikofer traveled through the dense Liberian forest in search of the famous and rare pygmy hippo (*Choeropsis liberiensis*), he was unable to find one himself. He therefore consulted local hunters for help and presented them a drawing of a hippo produced by his French colleague Alphonse Milne-Edwards (1835–1900), based on a brownish-reddish hippo skin held at Paris. Local West African informants, however, responded that they had never seen a hippo of such a color before.[1] In the nineteenth century, the conservation of hippo skins in European museums was difficult. Different institutions kept skins of different colors, leading to some confusion among biologists. Whereas they debated about animal skins rather than the observation of living hippos, Büttikofer, by contrast, relying on his networks of African hunters and informants, managed to collect not only carcasses, but information about pygmy hippo behavior and outward appearance. For instance, they were apparently black and a bit greenish, not brown or red.[2] The first person to publish these findings, however, was not the aspiring Swiss expeditioner himself, but the director of the Leiden Museum, Fredericus A. Jentink (1844–1913), who sponsored Büttikofer's second Liberian research trip from 1886 to 1887.

These transnational constellations that linked humans and nonhumans from around the world, characterized the reality of zoological and botanical collecting for European scientists and museums. The Dutch exploration of the tropical forest located in the de jure independent Republic of Liberia was for the main part an endeavor carried out by Swiss zoologists, botanists, and hunters with their West African research assistants. This chapter reconstructs how these explorations and networks shaped ethnological and zoological museums, research agendas, and zoos in Europe. It focuses on the scientist Büttikofer, the Swiss hunter Franz Xaver Stampfli, and their Liberian assistants from the Demery family, as they worked in and between Switzerland, the Netherlands, and Liberia.

Swiss and Liberian Villagers: Conceptualizing Scientific Networks from the "Peripheries"

Scientific networks in the colonial age were rarely organized simply according to the boundaries of nations or even empires. Complex social networks of states, villages, learned societies, museums, trade, and career and personal friendships overlapped. Historians have already employed networked approaches to understand the complexities of scientific endeavors in imperial contexts. Joseph Hodge, for example, argues for a networked conception of empire[3] for studies on the nexus of science and (British) imperialism. The network approach, he claims, would make it possible to better understand the hybrid nature of colonial scientific knowledge, drawing from indigenous informants and integrating both actors and ideas in "transnational, transregional, and transimperial" flows.[4] These concepts can be extended by new theories like Actor-network theory (ANT), which define not only non-European and European humans as part of the network, but animals and other nonhumans as actors in these networks. Although focusing on scientific practices in the colonial tropics, Warwick Anderson critiques ANT intellectuals like Latour who "managed to omit local agents and context, thus turning the network into a sort of iron cage through which no native can break. Thus, the 'local' seemed quite abstract, strangely depopulated, and depleted of historical and social content."[5] This chapter concentrates on a scientific network that conflated several imperial, national, and local structures by treating the so-called centers and peripheries as equals.[6] By integrating indigenous practices and ideas about animals in zoological-colonial encounters, and by conceiving animals as actors, the idea of a binary metropole-periphery concept is not a useful category.[7]

Encounters between Europeans of different nationalities were not uncommon in these networks. For much of the nineteenth century, European colonial governments hired scientists and technical experts from foreign countries in order to undertake research and management in forestry, botany, agriculture, veterinary science, and more.[8] Scientists not only used *one* "empire as a way to advance individual careers and secure state funding for research and institutional network-building"[9] but *various* empires and nation states. Since much global scientific networking was connected to the British Empire in the nineteenth century, it has received the most attention in current research.[10] Much less has been written about scientists from countries *without* colonies, who were employed by other colonial powers and utilized other imperial infrastructures, such as the Dutch or French, to establish themselves in transnational networks of science. Many of these scientists pursued work in conflict or concordance with their own national research agendas. Scientists often were employed by different colonial governments or institutions in the course of their careers, and they held ambivalent attitudes toward colonial policies. Some scientists actively distanced themselves from nationalist colonial agendas by attempting to pursue an academic approach emphasizing "universal" values in order to appeal to international audiences.[11] This fits within the wider trend of the professionalization of science, which emphasized values that transcended nationalism or place-based distinctions despite facing a rise of nationalism. Benedikt Stuchtey argues: "In sum, the connection between science and empire against the background of an international ethos, language, and universal standards on the one hand, and the establishment of the sciences within their national institutional frameworks on the other, did not necessarily form a contradiction."[12]

Against this background, the following case study sheds new light on important social and ecological aspects of zoological collecting. There is a growing recognition of the role of intermediaries and coproducers of scientific knowledge—especially in Human-Animal studies.[13] Historians of colonial science and exploration now argue that colonial zoological research tended to silence the complex social and ecological situations that led to the production of this knowledge. Instead of discussing how local hunters and mediators cooperated with European scientists, books and articles tended to offer charts, drawings, and other measurable, quantifiable data. Publications mostly neglected the contribution of indigenous experts, European missionary residents, or the often-ineffective hunting methods applied by the researchers from abroad. Animals, on the other hand, received considerable attention by colonial zoologists; but like other nonhumans, they were not regarded as agents or participants that actively shaped the research of the scientists. Most of the time, they appeared as still and dead objects for exposition until public zoos asked for live specimens for display and dissection. This chapter highlights the agency and role played by peoples from places once viewed as peripheral and also by nonhuman agents—animals in this case—in order to offer a more accurate historical picture of zoological collection and science in the age of high imperialism in Africa. Building on work by Hodge and Michael Worboys, the chapter demonstrates how exactly "the nature and content of the discipline itself was profoundly shaped by the imperial experience."[14]

Switzerland and the Dutch Colonial Empire in the Mid-Nineteenth Century: Entangled Colonial Research

Switzerland is a landlocked country that has since its creation lacked direct access to any ocean. In the nineteenth century, it was common for Swiss travelers to start their overseas trips from ships leaving Amsterdam.[15] Switzerland and the Netherlands had strong commercial, religious, and scientific ties during that century. This linkage via Dutch ports emerged as a result of the Swiss-Dutch trade in silk textiles and was further stimulated by the growth of Swiss Protestant missionary societies located at the Rhine River in Basel. Amsterdam, and the Netherlands, more generally, connected many Swiss people to the world outside of Europe.

Swiss naturalists, too, relied on their Dutch connections to procure zoological specimens for the growing number of natural history societies that developed in prosperous Swiss towns throughout the nineteenth century. The creation of these societies for natural history intensified contacts with colonial officials, researchers, and missionaries in order to enrich their collections of minerals, plants, and animals.[16] The Museum for Natural History in Leiden was, alongside other Dutch, German, and British institutions, a major partner in the intermuseum trade and exchange of such objects from European colonies. Being a rather shrinking colonial empire caused by the bankruptcy of the Dutch East and West India Companies and by the failing nationalization of their colonies, the Netherlands supported foreign research expeditions into all their dominions—a situation that has been described by Bernhard C. Schär as "paradoxical" and a characteristic of "crisis."[17]

The Dutch colonies were, first and foremost, part of an empire of trade that did not educate the necessary personnel to establish modern colonial statehood—that included colonial sciences at least since the late nineteenth century[18]—with regard to the British, French, and German models. Such bilateral contacts were embedded in a scientific environment that can be characterized as a "polycentric communications network"[19] challenging centralized regimes of colonial rule. Swiss scientists, in this context, acted in a transnational community of fellow researchers who were partners and competitors at the same time.[20] Switzerland, although never a colonial power, shared most of the important features of European colonial countries in the late nineteenth century.

It is widely recognized, though not always practiced, that historical studies of colonial contexts should address indigenous knowledge of colonial societies *and* European ideas of sciences.[21] It is, however, romantic to assume that indigenous knowledge has always been targeted at the preservation of biodiversity, just like it must be accounted for that colonial regimes were not mere protectors of nature.[22] Historians of science now retake into account places of science in a technical sense: laboratories, institutes, museums; whereas the fieldwork of zoologists is often marginalized.[23] If we ascertain the so-called scientific peripheries in the colonial world as important places for the production of knowledge about nature, then indigenous discourses about nature automatically become relevant features for the history of science, even without selecting an ethnobiological approach. This chapter addresses actors from different places and cultures of origin adequately, when their paths crossed in one colonial place: Liberia.

Between Fields and Museums: Zoological Expeditions and Career-Building in Colonial Liberia

Zoology emerged in the nineteenth century as a subdiscipline of biology. It was formed during the heyday of African colonization by European powers. In the 1880s, there still existed no fixed career path in zoology besides entering the profession as a taxidermist practitioner via natural history museums or universities. There was not even a clear line between botany and zoology within the natural sciences, although throughout the nineteenth century this line was slowly being drawn.[24] Exploratory journeys around the world included zoological research and publications on exotic fauna.[25] National categories were even less clear on these expeditions, as shown in the case of Swiss zoologists, who instrumentalized international colonial infrastructures in the West African Republic of Liberia in order to build up an international academic career. These zoologists came from a noncolonial state and traveled to an African state governed by black "repatriates" instead of a European colonial power.

The initiator of this zoological network was Johann Büttikofer from Switzerland. He was born in 1850 into a family of schoolteachers in the Emmental. Consequently, he was trained as a teacher himself and took up a post at a primary school in Grosswil. His intrinsic interest in zoology led him to assist the taxidermist of the Natural History Museum in Bern in his spare time. In 1876, he left his teaching position to become the taxidermist for the museum. Only two years later, and without any university education, he was given the position as an assistant at the Rijksmuseum van Natuurlijke Historie in Leiden (Netherlands), where he started his career as a scientist.

Swiss-Dutch exchange and communication between natural history museums had made this step possible. Büttikofer came from the petit bourgeois milieu in rural Switzerland and finished his career as the director of the Rotterdam Zoo. To gain this result, the Swiss zoologist had to walk the path of applied sciences in the taxidermist chambers and as a hunter abroad. With his reputation as a practitioner of zoology—rather than as scientific theorist—he managed to become, first, a conservator for dead animals in Leiden and, second, the keeper of living animals as director of the Diergaarde Blijdorp, the zoo of Rotterdam, from 1897 until his retirement and return to Bern.[26] Scientists, and even some historians of science, focused and privileged the laboratory over the field. The latter was therefore at first established as the workspace of the less educated zoologists.[27] Only by keeping animals in zoos did scientists start to look at animal behavior in their natural habitats and Büttikofer represents the establishment of zoos as new places of presentation and investigation of the exotic,[28] linking the metropolitan universities and the fields in the colonies. The Swiss researcher turned from a "skin zoologist" (*Balg-Zoologe*)[29] into a director of a zoo, whereas "the 'cathedrals of science' in the nineteenth century were not the zoos but the natural history museums."[30]

Büttikofer's first expedition to Liberia took place from 1879 until 1882. The independent African state of Liberia was dominated by black freed slave colonists from the United States. Its economy was driven by European trade factories along the coastline that purchased mainly palm oil, rubber, timber, coffee, and sugar from the interior. The Dutch factories supplied a minimum of European infrastructure that could be accessed during the expedition, and transport of animal exhibits was free for the expedition on all ships of one Dutch trade agency.[31] This factory was opened in 1855 and led by Hendrik Müller & Co. (Rotterdam) until it was taken over by the East African Company in 1892. When Büttikofer stayed in Liberia, the trade firm was running five factory stations in Liberia and was one of the three most powerful trade firms of that Republic, alongside E. Woermann (Hamburg) and Yates & Porterfield (New York).[32] The factory owners regarded themselves as patrons of scientific exploration. The Liberian government also supported foreign expeditions because neither Dutch nor Swiss actors followed the direct political interest of colonization in West Africa at that time.

The original route of Büttikofer was designed to follow the Atlantic coastline further east and south in order to explore the Gold Coast, Cameroun, and Angola rainforests over a period of six years.[33] Contacts were established with the Swiss staff of the Basel Protestant Mission in these destinations and regular shipping traffic along the African Atlantic coast was expected to guarantee passenger transport and fast shipping of organic exhibits that were vulnerable to decomposition and insect damage (especially by white ants) in the tropical climate of West Africa. Diplomats, merchants, and missionaries from different national origins supported the expeditions because scientific research was defined as a nonnationalist endeavor.

Büttikofer set out on his first journey to Liberia with his assistant Carl Friedrich Sala, a former colonial soldier from Dutch Java and Portuguese West Africa. Sala soon died from fever, and the Swiss zoologist also struggled with several attacks of disease and exhaustion. As a result, the expedition was stopped after two years, and Büttikofer was sent home to his Swiss village Inkwil to recover. It was in that period that the Natural History Society of the nearby town of Solothurn invited the explorer to give a lecture. He spoke about Liberia's politics, society, and nature. He expressed frustration in his lecture and lamented about "arrogant and lazy aborigines" who robbed and left the white travelers alone when they fell sick.[34] While slowly recovering in Switzerland, he must have developed the idea to send a representative to Liberia until his return. At the same time, he met the local pub owner and passionate hunter Franz Xaver Stampfli again.

Stampfli's background was different from that of Büttikofer. He was born in 1847 to a rather poor family, and following family traditions, he ran pubs with erratic success. He made plans to immigrate to the United States after running into financial difficulties. In 1882, Büttikofer offered Stampfli the opportunity to travel to Africa under the aegis of the Leiden Museum. Stampfli's wife passed away a year later in 1883, and shortly after her death, he finally decided to leave his children with relatives and accept Büttikofer's offer. Before starting his solo journey in 1884, he was trained as a taxidermist in Leiden for several months.[35] The museum wanted the Liberian expeditions to be strictly zoological. Stampfli, an experienced hunter, was thought of as the ideal assistant who could utilize his hunting experience from Switzerland when traveling in West Africa.

The two Swiss addressed each other as "friends" for a certain period.[36] When they were in Liberia in 1886 and 1887, however, Büttikofer became jealous of Stampfli's success in hunting, networking, and publishing his lay scientific results. Büttikofer instructed Stampfli to build up hunting stations in the hinterland while he liaised with other Europeans on the coast or recovered from his regular fevers. The two men rarely interacted or spent time together in Liberia. After only six months, Büttikofer returned to Leiden because of his ill health, leaving Stampfli in charge of hunting, buying, and dissecting animals for the museum.

Stampfli considered himself to be at the beginning of a successful career and hoped to gain a position at a museum. He even advised his son August in Switzerland to become a taxidermist, because he believed that they were required everywhere and would receive good salaries.[37] When becoming the head of the expedition, Stampfli started to collect animals for the museum of his hometown, Solothurn. He even donated the exhibits to them, not forgetting to mention that he would appreciate being given a job in the Solothurn Museum after his return from Liberia.[38] He regularly wrote letters to the Solothurn Society, where members read out his reports from West Africa and forwarded some of them for publication to the local newspaper *Solothurner Tagblatt*. In Büttikofer's eyes, however, Stampfli was not permitted to publish independently about the expedition. He contacted Solothurn from Leiden about this matter. Stampfli, still in Liberia, was upset about this and wrote to the *Naturforschende Gesellschaft Solothurn*: "Yesterday I received from my friends from the Kanton Solothurn the incredible news, that Büttikofer forbade you to publish my travel accounts. He is envious enough for such a statement to you. But he must not command me or you, the Natural History Society, in such matters."[39]

Stampfli may have anticipated that he could not win the competition for semiscientific positions in Leiden, where Büttikofer was already an established employee. Therefore, he tried to convince professional and lay scientists in his hometown to arrange a museum position for him—which he was in fact never offered. Moreover, he had to pay for his expenses in Africa (like Büttikofer) and was only refunded when he sent animal specimens to Leiden.[40] The two Swiss constantly derided each other in their letters: while Stampfli gossiped that his compatriot was sick or staying with friends most of the time, Büttikofer boasted that on his second journey he had stayed in a little hunting station "amongst the aborigines," whereas Stampfli had remained in a more metropolitan place.[41] Stampfli finally returned from his second trip to Liberia in 1888 and was never called again for any expedition or position from either Leiden or Solothurn. For some time, he still told people he might be part of a Dutch zoological expedition to Borneo. However, the Borneo expedition of the ethnographer Anton Willem Nieuwenhuis (1864–1953) set out with Büttikofer as the only zoological staff in 1893. The latter published a huge number of articles on that journey and was even granted an honorary doctoral degree by the University of Bern.

With his career in the gray area between craftsmanship and scholarship, and equipped with some academic prestige, Büttikofer became the director of the Rotterdam Zoo (1897–1924), while Stampfli tried in vain to get a post at the Solothurn Museum. Stampfli made a living as a taxidermist, stuffing animals for hobby hunters in Switzerland until his death in 1903. The obituaries published by his family nevertheless called him "Afrikajäger"[42] ("African Hunter") with some pride. Büttikofer spent the last three years of his life in Bern, and local oral stories have it that his family in Inkwil kept an exotic bird and a boy from his journey to Java in the 1890s. In Leiden, Büttikofer's work was well received. The popular Dutch version of the travel reports was brief and less detailed in comparison with the German volumes of *Reisebilder aus Liberia*. With his first voluminous report in his mother tongue, he clearly addressed a more scholarly readership, so that he abandoned the traditional order of travel reports arranged as chronological diary texts and instead organized the chapters by species: plants, animals, humans.[43]

Stampfli never published scientific articles or books. He hoped to be remembered by the species he discovered in West Africa, which were named after him. In this way, his last name was introduced into biological terminology. Most of his "discoveries" turned out to be only subspecies or had already been catalogued and named before. However, the flounder *Citharichthys stampflii* (Steindachner, 1894), for example, still bears the name of the Swiss hunter. For his solo journey (1884–86), Stampfli was given visual lists of animals already discovered, to help him to find new species. He observed carefully whenever an animal species was named after him in Leiden. During his second journey to West Africa, he noted the killings of animals already being called Stampfli: "Now the name Stampfli will never ever become extinct; specimens will be sent to America, Australia, Europe and to the greatest museums."[44] He would also ask the local newspapers to print notes and spread the news when the Leiden Museum announced another *stampflii* among their collections. However, the act of tracking and killing in Liberia did not entitle Stampfli to name and classify species. The scientists in Leiden decided whether a species was—often only valid for a short period—recognized as new. Despite Stampfli's ambitions, his activities in Liberia were not rewarded with a zoological career in the Netherlands, whereas Büttikofer established himself successfully in the Rotterdam Zoo at the edge of science and leisure culture.

Silencing or Educating African Zoologists

Many zoologists and geographers who participated in European exploration ignored names, stories, and methods of engaging with nature that African societies had developed before colonization. Büttikofer, for example, constantly complained about the malice of the climate and the unreliability and idleness of the "aborigines."[45] His assistant Stampfli, however, did not quite fit the mold of the scientific imperialist who despised all forms of indigenous knowledge. He differentiated between different groups in Liberian society and indigenous groups of the hinterland. He explained that he trusted the so-called natives, with whom he established intense contacts when staying in his remote hunting sheds, more than coastal Liberians from an immigrant background. Büttikofer, of course, engaged local servants and hunters to accompany him on his trips, to carry him in a hammock, or to supply him with various animals they had killed on their own. However, despite the huge amount of work done by Africans, his official words of thanks only included American and French missionaries, donors from Europe, and the staff of various trade factories. He neither mentioned his countryman Stampfli, nor any African servant, hunting assistant, or collector.[46] Instead, he blamed the porters and servants of the expeditions for minimizing the zoological results by their extreme noise of constant singing and talking.[47]

The Swiss zoologist was used to hunting quietly. African hunters in the east of Liberia, by contrast, motivated each other with songs when going into the forest equipped with weapons. This was particularly important when they hunted dangerous animals. When Büttikofer was informed by locals about a giant snake in the Liberian woods, the whole crowd accompanied the hunter with songs and later carried the killed animal into the village. There, local warriors repeatedly practiced mock attacks on the animal and destroyed its skin with their spears. Büttikofer was upset that he had successfully killed the giant snake but could not use the damaged skin for his collection. Instead, the local community cooked slices of the snake for a feast.[48] Communication between Büttikofer and local hunters often failed. Africans assumed the stranger was looking for "meat" and therefore killed and cooked the species required. This also happened in the case of pygmy hippos. Büttikofer could only buy the bones and teeth for his collection and not the skin he desired to have stuffed in Leiden.[49] Initially, the foreign traveler described his will to implement Swiss Alpine hunting practices and infrastructures in the Liberian mountainous landscapes; for instance, by constructing hunting huts.[50] However, during his trips in West Africa,

European hunting practices often failed, and local practices were either imitated or paid for as services.[51]

Soon after his arrival in Liberia, Büttikofer tried to access existing transcontinental networks of zoology by contacting the Liberian Mr. Lett, who previously had worked as a taxidermist for a German collector from Stettin. However, the zoologist from Leiden considered his services to be too expensive[52] and therefore engaged Jackson Demery from the Robertsport area. Demery and some other Vai hunters were paid as assistants during all the expeditions of Stampfli and Büttikofer. As a consequence, the African hunters from the coastal region engaged in the expeditions were themselves mostly hunting in unknown regions and unable to communicate with locals in their languages. Demery and his wife lived on an American Baptist missionary station close to Bendu.[53]

During the first trip, Demery hunted in the interior for the Leiden expedition while his daughter Mary collected insects and mollusks for Büttikofer.[54] Ties between Büttikofer and the Demery family were intensified when his son Archey Thomas Demery was sponsored from Leiden to stay at the museum for training as an animal preparator (November 1888 until August 1889). Back in Liberia, he was installed as a hunter and preparator to send specimens via the Dutch factory to Leiden. The strategy of Büttikofer was not to risk the lives of European zoologists to fevers of the tropical forests. Nevertheless, the young Thomas Demery died a year after his return to West Africa. Büttikofer wrote in his obituary: "This is a great loss for the sake of zoological investigation in Liberia and the neighbouring districts of Sierra Leone, and especially for our Museum which, by his death, loses its last direct connection with that part of Western Africa."[55]

Thomas Demery and his missionary family were the only Africans who were integrated into the zoological network as official assistants and who were deemed intellectually capable of generating zoological knowledge about Liberia and Sierra Leone. During his last hunting trip, Thomas Demery had crossed the border into the British colony of Sierra Leone, traveled up the Sulimah River, and shot various birds for Leiden. He probably died while the boxes of birds were still being shipped toward Europe. The Demerys had been picked as agents because the first European assistant of the expedition, Sala, had died, and Stampfli and Büttikofer also struggled with the threat of death several times. And yet, Stampfli was dismissed for other reasons and not to protect his European soul from African "fevers." Büttikofer seems to have suspected him of establishing contacts with other museums behind his back. At the same time, the young boy Demery was considered a civilized young scholar, equipped with European taxidermist knowledge. Despite ethnic and class differences, all three men went through the same form of education in the Leiden Museum preparatory section and were expected to work for a scientific museum.

The western concept of zoological knowledge at that time was still largely focused on the materiality of the animals: their skins, their size and weight, their inner organs. Therefore, Stampfli and Demery junior were asked to send in dissected animals, and only rarely living specimens. When zoology was linked to botany and mineralogy as well as to the natural history museum as its place of discourse, animals necessarily needed to be still and dead, and shipping live animals was economically risky.[56] Authenticity was only considered important with regard to the place where the animal body was killed and not to the hunter, his knowledge, or observations by locals. Büttikofer gave detailed explanations about the places and numbers in his scientific articles on Demery's specimens:

> After having spent nine months' time in our Museum, and thoroughly prepared and fit out to carry on my own investigations, Mr. A. T. Demery, the son of my Liberian huntsman Jackson Demery, left for his mother-country in August last year. Immediately after his arrival he went at work and has sent since, amongst many other objects, two small collections of birds from different parts of the district of Grand Cape Mount in Western Liberia. Many of the birds have been collected in the vicinity of Robertsport, others on the Johny Creek (a confluent of the Fisherman Lake), and others again at Jarjee, a Golah Town some days travel in the Interior up the Mahfa River. . . . The species, collected by us in Liberia, have now reached the number of 238.[57]

Demery was explicitly told to prepare the bodies of the birds and to send in lists of the places and days of killing. The focus was on the "production" of dead bodies and not on the behavior of the animals. The only honorable contribution of African and European assistants was to produce conservable exhibits on the ground. Indigenous knowledge about animal behavior was mostly silenced and ignored, if it did not directly lead to the killing, transportation, and conservation of the species.[58] The actual scientific part of the zoological work was to weigh, measure, and describe the bodies and their sex, and to give interpretations about their age based on outward appearance in the scientific museum space—the work of physiologists, anatomists, and taxonomists.[59] It was hence in the power of the museum zoologist to announce if new species had been discovered and to give them new names.

With each scientific step (tracking, killing, conservation, dissection, interpretation, publication) the professionalist, colonialist, and racist hierarchy of knowledge caused scientists to attribute an increasing portion of reputation to the actors.

Büttikofer honored the deceased Thomas Demery by calling a new species *Zosterops demeryi* and by explicitly stating that "I propose to name [it] after its discoverer."[60] This so-called African Yellow White-eye bird is still known as *Zosterops senegalensis demeryi* today. Thomas Demery—like Stampfli—was accepted as a discoverer of new species because of his training abroad, whereas other African assistants and servants who explained or sold animals to Büttikofer were rarely mentioned and certainly never commemorated with names.[61] This means that a discovery of a new species in that sense required its killing *and* preparation along European taxidermist methods.

Stampfli and Demery senior often went hunting as a team. Although they were experienced hunters, they often failed with regard to big game—the most prestigious prey. Büttikofer therefore concluded that "one must send native hunters" for such complicated work and argued that the black skin of the "natives" allowed them to approach game more easily and quietly.[62] This clearly contradicts former ideas about "native Africans" being too noisy for successful hunting expeditions (see above). After having changed his mind about African hunting practices, Büttikofer was more open to cooperating with local hunters on his second trip. He would wait for locals to kill a hippo and then send Demery to assist the people slaughtering it in the way that skins and bones could be used for the zoological collection and the flesh was still cooked and smoked by the local population.[63] Such win-win situations were typical for colonial big-game hunting all over Africa,[64] and it took Büttikofer a rather long time to realize the efficiency of integrating local hunters into his networks. He began to explain to them which species he was interested in and followed the cultural etiquette more carefully by sharing spirits—"the hunter's drink"[65]—after paying them in kind. His local counterparts, on the other hand, highlighted Büttikofer's work as a taxidermist, more than as an actual hunter, by calling him "animal-skinner" in their language. Another description emphasized his specific interest in birds by referring to him as the "bird-killer."[66] In this way, they depicted a hunting practice that was untypical in the Liberian hinterland subsistence economy.

European zoologists and coastal Liberians were generally doubtful about the hunting skills and zoological knowledge of locals. They never deemed their experiences with animals worth dealing with scientifically. Information on animal behavior was either considered important with regard to successful hunting and trapping, or it was regarded as entertaining anecdotes. However clearly this hierarchy of expertise was presented at first glance, the manifold actors reveal an embattled and undefined social scheme: a nonacademic hunting assistant from Switzerland, a black missionary educated in taxidermy in Leiden, local hunters becoming involved as distributers of animals. On the one hand, scientific hierarchies were built upon racist concepts or ideas about class; on the other hand, however, the newborn profession of "zoologists" was still vague and in the making. The production of knowledge about West African wildlife therefore included various actors trying to make a career in Europe by their occupations in Africa. Some of them were silenced and failed, some of them succeeded.

The Closest Animals: Trading and Taming

For Europeans residing in Liberia, there were certain iconic local species they would usually talk about: the small West African "pygmy" hippo, the manatees in the rivers, and chimpanzees. All three species are mammals and at least two of them — manatees and apes — had been historically discussed as mythic animals or human-animal hybrids in North Africa and Europe.[67] Within West Africa, manatees were often referred to as half-human cannibals,[68] whereas in European discourses they were long identified as sea cow mermaids. Chimpanzees, on the other hand, often interacted with people more directly.[69] The idea of conceptualizing chimpanzees as especially close relatives of the human race was in fact not a European invention. Usually, monkey species were killed and eaten by local communities of the Liberian and Sierra Leonian hinterland, except for chimpanzees: "Because they were too much like man."[70] An old man allegedly told Büttikofer that they would even imitate the fire places of the humans — though without the fire.[71] Their social behavior was considered to be modeled after human social structures. While many West African societies deduced a general taboo on eating chimpanzees from this resemblance, the Swiss hunters killed, cooked, dissected, and ate chimpanzees during their expeditions. Büttikofer mentioned no moral doubts about this diet, but he described that it was hard to stand a chimpanzee looking at you while killing it. The legitimacy of this act, he explained, was that this was an "animal sacrifice for science."[72] When he injured an ape with his rifle and took over thirty minutes to remove the dying animal from a tree, the Leiden zoologist refused to describe the horrible scene and from that time on promised never to kill an ape again.[73] Such spiritual experiences of horror and enlightenment while witnessing the suffering of animals often served as propagandistic narratives to illustrate how individuals changed from big-game hunters to environmentalists and animal rights activists. Turns like this were typical for post–1900 Western hunters in Africa. Büttikofer developed a sincere sense for the protection of animals after his return to the Netherlands, where he was engaged in the Dutch Society for the Protection of Birds (1909–24).[74] When in Africa, however, Büttikofer regularly killed animals at a certain locality until no more of them were found.[75]

After Büttikofer himself had already stopped killing chimpanzees and focused on caging those given to him as presents or trade items, he still accepted chimpanzee skulls prepared by the Liberian missionary Jackson Demery.[76] His own initiatives, at this time, were centered on his African menagerie of living animals. His Liberian expeditions included the transport of some living animals to Europe, which often meant that animals were caged at the stations of the Swiss hunters and died before they left the continent.[77] Throughout the voluminous travel accounts of Büttikofer, the goal of these caging practices remains obscure because no scientific reports or results relied on the animals' imprisonment or observation. The scientific purpose of caging animals was rather nebulous. Furthermore, contacts with zoos had not been established by the Leiden Museum before the expeditions. Here, the role of the European trade agents living on the Liberian coast has to be taken into account. Büttikofer and Stampfli did not start collecting living animals themselves, but received tamed chimpanzees and a tamed red river hog from a Belgian trade agent as gifts.[78] The hog and a chimpanzee had lived at the factory and reportedly were playmates until Büttikofer had them shipped to Europe and given to the Amsterdam Zoological Garden, which marked the start of Büttikofer's engagement with public zoos.[79]

Keeping and taming exotic animals were common practices of foreign merchants in West Africa, who inspired and motivated the zoologists to cage and trade them. Using exotic African animals as intercultural diplomatic gifts has a long history reaching back to ancient Egyptian trade networks. They were used as symbols of power over nature and strange countries, and could serve as signs of the superiority of the colonial or Western self in opposition to African societies and natures.[80] By exchanging living animals as a method of self-affirmation, Belgian, French, or German traders built social bonds with the Swiss zoologists. Other than European trade agents who had been stationed in the area for years or decades, Büttikofer and Stampfli were newcomers without a fixed milieu they naturally belonged to—they were neither missionaries, nor colonial or diplomatic staff, nor merchants. In this unclear situation, they soon socialized with European agents and copied their lifestyle to a certain extent. Encouraged by the animal gifts, they started to buy living animals systematically from local traders. They purchased snakes, palm civets, and other species instead of trapping them on their own,[81] because they lacked the skills to catch them unharmed. This specialization in living animals turned the zoological staff into part-time traders and Büttikofer into a zoo director, some ten years after his African journeys.

The collection of animals dead or alive and the provisional placing of them in small cages and boxes caused various problems because of the organic materials. The dried skins attracted big ants that destroyed prepared animals and managed to kill some smaller ones.[82] One of the monkeys of the menagerie once took a prepared bird from a box and brought it to the hunting dog, which ate it and nearly died from the arsenic.[83] Büttikofer, Stampfli, and Demery were unprepared for keeping such a collection of living and formerly living beings. Their education included killing and preparing, not feeding and caging of animals. During their travels, the observation of animal behavior was expressed in anecdotal ways and not within scientific sections of the travel report or in individual articles. The chimpanzees they kept at the hunting stations were characterized by their vivid interaction with humans—aggressive, loving, or tricky—and not with any reference to their "natural" behavior with their conspecifics. Büttikofer explained their acting in human terms, narrating how he pretended being dead, stopped breathing, and was amused about his monkey reacting with shock and opening his eyes softly with his fingers to bring him back to life.[84] The contradiction between the killing of some apes and sensitive interaction with tamed chimpanzees is obviously extreme.

While European trade staff and missionaries kept chimpanzees as pets, local societies largely abstained from both eating and taming them. They would sometimes catch the babies to trade them further to the coast where they were used as pets or workers and not to stay in their homes of the hinterland.[85] Chimpanzees were considered as potentially aggressive.[86] The Liberian hunter Demery once experienced the dangers of attacking chimpanzee groups. When he shot an individual, the group did not escape but started a counterattack and injured him severely. In the end, he was not even able to confiscate the dead ape for Büttikofer's collection because the deceased was evacuated and carried away by its group.[87]

In the border region of Sierra Leone and Liberia, attacks of chimpanzees on humans have been frequent since at least the late nineteenth century. In the British Protectorate of Sierra Leone they were attributed to acts of so-called human-baboon transformations. The idea that sudden deaths by diseases or drowning and by attacks of wild animals were caused by "human-animals" had been present in West Africa for a long time. Locally, individuals or groups were tried and executed for turning into a beast and killing humans, until British colonial law attempted to integrate these accusations into their legal framework.[88] From the late 1880s until 1941, numerous colonial trials were held dealing with human-animal murders. During the First World War, stories about apes and human-baboons killing individuals created some local panic, so that W. B. Stanley, a colonial district commissioner and passionate elephant hunter, published a semiscientific article about the question of "Carnivorous Apes in Sierra Leone."[89] Although admitting in this text that many West Africans told stories about real chimpanzees attacking and killing humans, Stanley concluded that "it is important to remember that these cases of injuries and death are regarded by the European and the native from completely different standpoints. The European view entirely precludes the idea of witchcraft and dismisses the idea of real apes as being impossible; it concludes, therefore, that the tragedy must be the work of human beings."[90]

Although Stanley thought that the killings were done by human beings, he mentioned some six cases when chimpanzee groups or male individuals actually attacked humans, occupied their farm sheds, or even took away and killed babies.[91] In the years after this publication he started to believe in the theory that all "human-baboon" attacks were caused by chimpanzees. He had chimpanzee traps installed in his district and presented the corpses to the village public in order to teach them that these animals were not humans or bewitched by humans. He narrated these killings as a success of colonial education of superstitious Africans leading them toward Western rationality, when it was in fact the British colonial government and himself that had supported the idea of human-animal hybrids before: "The medicine men were very sick, and we have been the best of friends with the natives ever since. The exploding of the ape-man superstition did it all."[92]

Back in the 1880s, Büttikofer's knowledge about primate behavior was mainly linked to his experiences as a hunter until he started experimenting with bringing up orphaned ape infants. Büttikofer tried to nourish the monkey babies with bananas and condensed milk—without success. Although not used to interspecies adoption, a local chief offered the visitor a little monkey breastfed by a young mother of the village.[93] She was probably paid for this service to carry the monkey together with her own baby on her back like twins. The little monkey survived by this treatment until Büttikofer took it away from the woman to the capital Monrovia. What the European zoologist observed as an indigenous practice, was quite the opposite: this adoption was a reaction to Büttikofer's presence because tamed animals were not kept in the Liberian hinterland at all.[94]

Taming measures in colonial Africa demonstrated the racist attitude of the colonizers.[95] African humans and animals alike were conceptualized as imperfect and raw beings that had to be educated, civilized, and disciplined. To mark the "Africanness" of Liberian animals and the European nature of his hunting dog, Büttikofer used human names to draw lines among the animals of his menagerie. Hence his Swiss hunting dog was named "Bello"—a typical name for dogs in German-speaking Europe—whereas one of his monkeys was called "Jack" after the nickname of his African hunting assistant Jackson Demery. Moreover, the red river hog later given to the Amsterdam Zoo was named Jassa after his African housekeeper in Hill Town. And with a paternalistic and ironic tone, he added that the hog was always "well-behaved."[96] Colonialist cultures of taming and pet-keeping in Liberia were developed by European agents within the compounds of their trade factories. It is likely that they also introduced the names of African servants for these animals. Companionship with (formerly) wild animals was unknown to indigenous cultures of the region, in which wildlife was excluded practically and spiritually from human spheres.[97] The Swiss zoologists who had initially arrived as taxidermists researching anatomy, slowly adapted to the international colonial expatriate lifestyle, and this included the taming of animals. By this step, they almost unintentionally became "experts" on exotic living beings and bridged the gap between museum and zoo businesses.

Conclusion

In the nineteenth century, zoologist ideas about scientific exploitation and an all-embracing classification of African wildlife determined the way zoological networks were arranged: animals generally had to be killed or "sacrificed" for science in order to become a still object of research and evidence for the fieldwork of hunters and gatherers of new species. Local hunters and traders were integrated into these networks, if required, and could even become honored members by being educated by Europeans (in Europe). It was manifest, however, that Africa remained their domain of scientific assistance and not the zoological museums in the West. Like the animals of science, the human staff was expected to dedicate health and often to sacrifice lives to accomplish their field work in tropical forests. Those men (and rarely women) were engaged as hunters, gatherers, collectors, zoologists, animal keepers, and traders at the same time. Professional and lay staff combined several professions—they could be traders and guides, missionaries and hunters, scientists and collectors. In the heyday of colonialism, the collection of killed animals and the trade and exposition of those still alive were closely entangled. Furthermore, zoography and ethnography were neither separate in academic organization nor in terms of the places of exhibition.

In this complex intercultural setting of zoological research, new professions, scientific disciplines, networks, and species were created in a process of knowledge production and transfer. Often, European scientific ideas and hunting practices failed in the Liberian forests; therefore, the skills of local hunters were required and paid for. The Swiss zoologists tried to educate Christian Africans as intermediaries to use Western hunting and taxidermist methods. At the end of the day, however, big mammals were shot or caught more effectively by hunters using indigenous instruments and tactics. At this point, the transfer of knowledge was blocked: the foreign zoologists probably were not interested in acknowledging local wisdom about certain species, their habits, habitats, and cultural significance in local cultures. They used fieldwork as a way to gain animal bodies, and their biological approach relied on a materialist and anatomist research agenda. Given the fact that indigenous Liberian cultures considered certain wild mammals the source of male political, economic, and sexual power, they may well have prevented strangers from being introduced into their world of wild animals. What we often interpret as European ignorance of African animal knowledge out of arrogance must in fact also be approached as ignorance caused by the refusal of African experts to share what they knew. Hierarchies were more complex than the mere dominance of Europeans over Africans.

In this environment, intermediary traders delivered animals for zoological use, whereas local knowledge about them was mostly lost. The foreign hunters from the Liberian coast or Europe sometimes exchanged stories about the animals with indigenous hunters, but while local culture focused on the "characters" of wild animals using a universalist approach, Western scientific culture only credited the dissection of their bodies and functions. Revisiting the debate of "pygmy" hippos, the Swiss Büttikofer referred to local informants for clarification on the color of these animals, whereas he completely copied statements on their behavior from Western armchair naturalists. He picked up the idea of anatomical similarities between hippos and wild pigs expressed by predecessors and included similar behavioral traits in this theory: The hippos would spread over huge habitats, migrate constantly, and taste just like wild pigs.[98] However, this reproduced Western hierarchy of animal knowledge collapsed when Büttikofer was forced to explicate the animals he was looking for in order to buy them from local hunters. In the contact zone, the hierarchies of imperial science were constantly contested, and narrative descriptions of animals were challenged. Whether a pygmy hippo appeared black, green, or red depended on the context of observation—as a dried skin, as a cadaver, in a forest, in water, or covered with their own pinkish secretion to sun-protect their skin. What was accepted as scientific truth and in what way was a matter of authority and network access.

Notes

1. Fredericus A. Jentink, "Zoological Researches in Liberia: A List of Mammals, Collected by J. Büttikofer, C. F. Sala and F. X. Stampfli, with Biological Observations," *Notes of the Leyden Museum* 10 (1888): 31.

2. Jentink, "Zoological Researches in Liberia," 30.

3. Joseph M. Hodge, "Science and Empire: An Overview of the Historical Scholarship," in *Science and Empire: Knowledge and Networks of Science across the British Empire, 1800–1970*, edited by Brett M. Bennett and Joseph M. Hodge (Basingstoke, UK: Palgrave Macmillan, 2011), 3.

4. Hodge, "Science and Empire," 19.

5. Warwick Anderson, "Remembering the Spread of Western Science," *Historical Records of Australian Science* 29 (2018): 77. Likewise have scientific servants and "invisible technicians" long been ignored across the history of sciences. An exception to this shortcoming is Steven Shapin, *A Social History of Truth: Civility and Science in Seventeenth-Century England* (Chicago: University of Chicago Press, 1993), especially chapter 7.

6. In web-like networks, a periphery may also become a metropole crossing the borders of empires. For a discussion, see Hodge, "Science and Empire," 17. Michael Chayut discussed scientific peripheries as the places and spheres of innovation (Michael Chayut, "The Hybridisation of Scientific Roles and Ideas in the Context of Centres and Peripheries," *Minerva* 32 [1994]: 297–308).

7. See Roy MacLeod, "Introduction," in *Osiris*, vol. 15, *Nature and Empire: Science and the Colonial Enterprise* (Chicago: University of Chicago Press, 2000), 6. However, MacLeod's earlier nomenclature of "imperial" and "colonial science" was based on the dichotomous idea of a—although potentially moving—metropolis and peripheries. Cf. "On Visiting the 'Moving Metropolis': Reflections on the Architecture of Imperial Science," in *Scientific Colonialism: A Cross-Cultural Comparison*, edited by Nathan Reingold and Marc Rothenberg (Washington, DC: Smithsonian Institution Press, 1987). At least since the 1980s, science was deemed a polycentric network and Anderson gives an excellent historical outline of MacLeod's, Chambers's and Gillespie's responses to Basalla; Anderson, "Remembering the Spread of Western Science," 75–76.

8. See, for example, Ulrike Kirchberger, "German Scientists in the Indian Forest Service: A German Contribution to the Raj?," *Journal of Imperial and Commonwealth History* 29 (2001): 1–26.

9. Hodge, "Science and Empire," 14. In this section he discusses the work of Michael Worboys.

10. Brett M. Bennett, "The Consolidation and Reconfiguration of 'British' Networks of Science, 1800–1970," in *Science and Empire,* 31.

11. The German geographer Heinrich Barth, for example, led several African expeditions for the British government and the Royal Geographical Society in the 1850s. Barth's cartographies were used for colonization, but he often distanced himself from colonial policies as a scholar. His publications were sponsored by the British government, but he criticized their politics in Africa; see Heinrich Schiffers, "Heinrich Barths Lebensweg," in *Heinrich Barth: Ein Forscher in Afrika,* edited by Heinrich Schiffers (Wiesbaden, Germany: Franz Steiner, 1967), 51. Like Barth, the Swiss zoologists in this chapter identified with certain scientific institutions (museums, learned societies) but less so with (Dutch) colonialism. These scientists often published in languages other than those spoken in the country they worked for. Stuchtey even attributes characteristics of "fraternities" to international scientific networks in the age of modern colonialism (Benedikt Stuchtey, "Introduction: Towards a Comparative History of Science and Tropical Medicine in Imperial Cultures since 1800," in *Science across the European Empires, 1800–1950,* edited by Benedickt Stuchtey [Oxford: Oxford University Press, 2005], 15).

12. Stuchtey, *Science across the European Empires,* "Introduction," 45.

13. Research projects on the history of zoos, for instance, came to highlight the role of zookeepers as intermediaries between animals and scientists in the production of knowledge about animals. For a research agenda on the history of science and animals, see Mitchell G. Ash, "Tiere und Wissenschaft: Versachlichung und Vermenschlichung im Widerstreit," in *Tiere und Geschichte: Konturen einer Animate History,* edited by Gesine Krüger, Aline Steinbrecher, and Clemens Wischermann (Stuttgart: Franz Steiner, 2014), 290.

14. See Hodge's remarks on Michael Worboys ("Science and Empire," 19).

15. Bernhard C. Schär, *Tropenliebe: Schweizer Naturforscher und niederländischer Imperialismus in Südostasien um 1900* (Frankfurt: Campus Verlag, 2015), 61.

16. Schär, *Tropenliebe,* 88.

17. Schär, *Tropenliebe,* 98, 331, 333. Florian Wagner's chapter in this volume also highlights the colonial Dutch internationalization strategy.

18. In special journal issues or anthologies on colonial science, articles about Dutch colonies are often missing. Cf. MacLeod, "Introduction," 13. For agricultural research in Dutch Indonesia, see Suzanne M. Moon, "Development, Technology, and the Unique Economy of the Colony: The Dual Economy Thesis in Netherlands East Indies' Development Policies, c. 1920," in *Science across the European Empires.*

19. David Wade Chambers and Richard Gillespie, "Locality in the History of Science: Colonial Science, Technoscience, and Indigenous Knowledge," in *Osiris,* 15:223. The authors highlight that the ancient Greek and modern Western idea of scientific knowledge was not bound to a certain locality, whereas colonial science has in fact always been shaped by places and distances (15:228). The colonial places of research—except the more or less private zoos of colonial officials in the capitals—became more relevant only with the rise of ethology (study of animal behavior) in the twentieth century; see Richard W. Burkhardt Jr., "Ethology, Natural History, the Life

Sciences, and the Problem of Space," *Journal of the History of Biology* 32 (1999): 490; for a more general discussion of the spatial factor in biology, see Diarmid A. Finnega, "The Spatial Turn: Geographical Approaches in the History of Science," *Journal of the History of Biology* 41 (2008).

20. Schär, *Tropenliebe*, 125.

21. Wade Chambers and Gillespie, "Locality in the History of Science," 229. However, instead of speaking of a general "subjugation" of local knowledge, I assume a massive European ignorance about indigenous ideas about nature.

22. Wade Chambers and Gillespie, "Locality in the History of Science," 233.

23. For the changing boundaries between field and laboratory biologists, see Robert E. Kohler, *Landscapes and Labscapes: Exploring the Lab-Field Border in Biology* (Chicago: University of Chicago Press, 2002).

24. William Coleman, *Biology in the Nineteenth Century: Problems of Form, Function and Transformation* (Cambridge: Cambridge University Press, 1977).

25. Charles Darwin, *The Zoology of the Voyage of H.M.S. Beagle Under the Command of Captain Fitzroy, R.N., during the Years 1832 to 1836: Published with the approval of the Lords Commissioners of Her Majesty's Treasury*, 5 vols. (London: Smith, Elder & Co. 1838–43). Although Darwin highlighted the necessity of observing animals in their natural environments, his publications did not result in a rise of field studies, but in the heyday of anatomic research in animals. See Burkhardt, "Ethology, Natural History, the Life Sciences, and the Problem of Space," 494.

26. Lipke Bijdeley Holthuis, *1820–1958: Rijksmuseum van Natuurlijke Historie* (Leiden: Karstens Drukkers 1995), 58–59.

27. "Observation and comparison, once performed by closet and voyager naturalists alike, became second-best practices in a landscape dominated by labs." Kohler, *Landscapes and Labscapes*, 3.

28. At first, "the interest of the zoologists and morphologists lay primarily in the carcasses provided by the zoo" (Oliver Hochadel, "Science in the 19th-Century Zoo," *Endeavour* 29 [2005]: 41) because "the observation of living animals was simply not on university curricula and research agendas" (p. 39) whereas they established themselves as places for studies on animal behavior when zoo biology "became a discipline of its own from the 1930s onwards" (p. 40).

29. As early as 1871, the taxidermist approach of biologists was ridiculed by this term. See Ernst Friedel, "Thierleben und Thierpflege in Holland, England u. Belgien," *Der zoologische Garten* 12 (1872): 331.

30. Hochadel, "Science," 39. See also Susan Sheets-Pyenson, *Cathedrals of Science: The Development of Colonial Natural History Museums during the Late Nineteenth Century* (Kingston: McGill-Queen's University Press, 1988). Sheets-Pyenson refers the emergence of these "cathedrals" to nationalist agendas.

31. Johann Büttikofer, *Reisebilder aus Liberia: Resultate geographischer, naturwissenschaftlicher und ethnographischer Untersuchungen während der Jahre 1879–1882 und 1886–1887*, 2 vols. (Leiden: Brill, 1890), 1:vi.

32. Johann Büttikofer, "Zustände in der Negerrepublik Liberia," *Deutsche Geographische Blätter* 7 (1884): 86.

33. Büttikofer, *Reisebilder aus Liberia*, 1:v.

34. Protokoll Naturforschende Gesellschaft Solothurn, entry of the 11.12.1882, Archive of the Naturmuseum Solothurn.

35. Hans R. Stampfli, "Der Afrikajäger Franz Xaver Stampfli (1847–1903): Ein Solothurner entdeckt neue Tierarten und bereichert naturwissenschaftliche Sammlungen," *Mitteilungen der Naturforschenden Gesellschaft des Kantons Solothurn* 36 (1993): 123–35.

36. Franz Xaver Stampfli, "Afrikanische Bilder: Nach den brieflichen Mittheilungen eines solothurnischen Landsmannes," *Solothurner Tagblatt*, 29 May 1885, 3.

37. *Solothurner Tagblatt*, 31 May 1885, 3.

38. Hans R. Stampfli, "Der Afrikajäger Franz Xaver Stampfli," 128.

39. "Habe gestern von meinen Freunden aus Kt. Bern die unglaubliche Nachricht erhalten, dass Büttikofer Ihnen soll verboten haben, meine Reiseberichte zu veröffentlichen. Er ist neidisch genug eine solche Aeusserung an Sie gemacht zu haben. Er hat aber in solchen Sachen weder mir noch Ihnen, und der N'forschenden Gesellschaft zu befehlen." Letter from Stampfli, 12 October 1887, Zentralbibliothek Solothurn.

40. Franz Xaver Stampfli, "Afrikanische Bilder," 31 May 1885, 3.

41. "Mitten unter den Eingebornen." Johann Büttikofer, "Brief des Herrn J. Büttikofer," *Jahresbericht der Geographischen Gesellschaft von Bern* 8 (1885–1887): 78.

42. Lina Stampfli, Todesanzeige, 17 July 1903, Correspondenzen 1903, Naturmuseum Solothurn.

43. In 2013, Brill published an annotated English translation of his major work in German: Henk Dop and Phillip Robinson, eds., *Travel Sketches from Liberia: Johann Büttikofer's 19th Century Rainforest Explorations in West Africa* (Leiden: Brill, 2013), ix.

44. "Nun stirbt der Name Stampfli nie und nimmer aus; die Exemplare werden nach Amerika, Australien, Europa u. an die größten Museen versandt." Franz Xaver Stampfli, "Zweiter Reisebericht von Xaver Stampfli in Liberia," *Solothurner Tagblatt*, 7 January 1888, 2.

45. Büttikofer, *Reisebilder aus Liberia*, 1:vii.

46. Büttikofer, *Reisebilder aus Liberia*, 1:xii.

47. Büttikofer, *Reisebilder aus Liberia*, 1:31.

48. Büttikofer, *Reisebilder aus Liberia*, 1:122–23.

49. Büttikofer, *Reisebilder aus Liberia*, 1:138, 153.

50. Büttikofer, *Reisebilder aus Liberia*, 1:99, 108. For a discussion of the Swiss affinity for mountainous African regions, see Lukas Zürcher, "So fanden wir auf der Karte diesen kleinen Staat: Globale Positionierung und lokale Entwicklungsfantasien der Schweiz in Ruanda in den sechziger Jahren," in *Entwicklungswelten: Globalgeschichte der Entwicklungszusammenarbeit*, edited by Hubertus Büschel and Daniel Speich (Frankfurt: Campus, 2009).

51. Büttikofer, *Reisebilder aus Liberia*, 1:125.

52. Büttikofer, *Reisebilder aus Liberia*, 1:318.

53. Büttikofer, *Reisebilder aus Liberia*, 1:286, 288.

54. Büttikofer, *Reisebilder aus Liberia* 1:292.

55. Johann Büttikofer, "Archey Thomas Demery: Obituary," *Notes from the Leiden Museum* 13 (1891): 248.

56. See Burkhardt, "Ethology, Natural History, the Life Sciences, and the Problem of Space," 490. Büttikofer combined both identities, that of a "naturalist voyager" (or "field naturalist") and a "museum naturalist."

57. Johann Büttikofer, "Zoological Researches in Liberia: On a Series of Birds, collected by Mr. A. T. Demery in the District of Grand Cape Mount," *Notes from the Leyden Museum* 12 (1890): 197–98.

58. For the exclusion of experts based on race, class, and gender, also see Carey MacCormack's chapter.

59. This early period of the history of biology was linked to zoological sections of natural history museums and to the so-called naturalists' work. See Coleman, *Biology in the Nineteenth Century*, 2–11.

60. Büttikofer, "Zoological Researches in Liberia: On a Series of Birds," 202.

61. Carey MacCormack's chapter describes the same strategic omission of non-European names.

62. Büttikofer, *Reisebilder aus Liberia*, 1:324.

63. Büttikofer, *Reisebilder aus Liberia*, 1:346–47.

64. John M. MacKenzie, *The Empire of Nature: Hunting, Conservation and British Imperialism* (Manchester, UK: Manchester University Press, 1997), 57.

65. Büttikofer, *Reisebilder aus Liberia*, 1:339.

66. Büttikofer, *Reisebilder aus Liberia*, 2:297.

67. Büttikofer, *Reisebilder aus Liberia*, 1:271.

68. Stephanie Zehnle, "A Geography of Jihad: Jihadist Concepts of Space and Sokoto Warfare (West Africa, ca. 1800–1850)" (PhD diss., University of Kassel, Germany, 2015), 394.

69. The biological nomenclature was still rudimentary in the 1880s, so that this primate species was usually called "black baboons" or "big black ape." Cf. Büttikofer, *Reisebilder aus Liberia*, 1:76.

70. Büttikofer, *Reisebilder aus Liberia*, 1:229.

71. Büttikofer, *Reisebilder aus Liberia*, 1:230.

72. Büttikofer, *Reisebilder aus Liberia*, 1:237.

73. Büttikofer, *Reisebilder aus Liberia*, 1:294.

74. After his death, an ornithological reserve on the island of Texel was named "Büttikofer's Mieland."

75. Büttikofer, *Reisebilder aus Liberia*, 1:126.

76. Büttikofer, *Reisebilder aus Liberia*, 1:297.

77. Büttikofer, *Reisebilder aus Liberia*, 1:112, 382.

78. Büttikofer, *Reisebilder aus Liberia*, 1:417.

79. Büttikofer, *Reisebilder aus Liberia*, 1:436.

80. Stephanie Zehnle, "War and Wilderness: The Sokoto Jihad and Its Animal Discourse," *Critical African Studies* 8 (2016): 216–37.

81. Büttikofer, *Reisebilder aus Liberia*, 1:366.

82. Büttikofer, *Reisebilder aus Liberia*, 1:383.

83. Büttikofer, *Reisebilder aus Liberia*, 1:384.

84. Büttikofer, *Reisebilder aus Liberia*, 1:385.

85. Büttikofer, *Reisebilder aus Liberia*, 2:350.

86. Chimpanzee attacks are still problematic in rural West African regions; see Kimberley J. Hockings, Gen Yamakoshi, Asami Kabasawa, and Tetsuro Matsuzawa, "Attacks on Local Persons by Chimpanzees in Bossou, Republic of Guinea: Long-Term Perspectives," *American Journal of Primatology* 72 (2010): 887–96.

87. Büttikofer, *Reisebilder aus Liberia*, 1:257.

88. Stephanie Zehnle, "Of Leopards and Lesser Animals: Trials and Tribulations of the 'Human-Leopard Murders' in Colonial Africa," in *The Historical Animal*, edited by Susan Nance (New York: Syracuse University Press, 2015), 221–38.

89. W. B. Stanley, "Carnivorous Apes in Sierra Leone," *Sierra Leone Studies* 2 (1919): 3–19.

90. W. B. Stanley, "Carnivorous Apes," 5.

91. The same can be observed in regional newspapers that simultaneously reported about the "Human Baboon Society" killing four persons (Macaia, "Protectorate Affairs," *Sierra Leone Weekly News* [9 November 1940]: 4, 9); and about real "Baboons" attacking girls on a farm (*Sierra Leone Weekly News* [27 September 1941]: 5, 8).

92. Anonymous, "Sierra Leone Terrorized by Apes: British Hunters End Killings Charged to Bewitched Animals," *Cambridge Sentinel*, 31 January 1925, 3.

93. Büttikofer, *Reisebilder aus Liberia*, 2:355.

94. Local populations in Liberia and Sierra Leone even resisted the introduction of cows. See W. B. Stanley, "Sherbro District Report 1916," The National Archives, Kew (London), CO 267/575, 20 May 1917.

95. Bernhard Gissibl, "Das kolonisierte Tier," *Werkstatt Geschichte* 56 (2010): 7–28.

96. Büttikofer, *Reisebilder aus Liberia*, 1:384; 2:388. A comparative analysis of colonialist name-giving cultural practices for pets and tamed wild animals could be enlightening but has not been attempted so far.

97. Zehnle, "Of Leopards and Lesser Animals."

98. Jentink, "Zoological Researches in Liberia," 30.

Adapting to Change in Australian Estuaries

Oysters in the Techno-Fix Cycles of Colonial Capitalism

JODI FRAWLEY

Settlers arriving in New South Wales over the nineteenth century found the estuaries unimaginably dense with life. They remain complex ecological zones between the ocean and the inland, amid saltwater and fresh, and provided a network of wetlands stretching from Eden in the south to the Tweed River in the north. Each of the 130 estuaries in New South Wales is unique, depending on the relationship between the vegetation and the aquatic and geomorphic features. Some have wide bays, others a series of interlinking lakes; some are dense with mangroves or salt marshes or wallum forests. Some have all three. Some are hugged by towering sandstone outcrops while others interact with constantly changing dune activity. P. S. Roy and others classified four types of estuaries in Eastern Australia: open ocean embayment, drowned river valley estuaries, barrier estuaries, and intermittent (saline coastal lakes) estuaries.[1] All provided for people, plants, and animals that belonged to one another for at least 65,000 years.[2]

All estuaries supported populations of oysters. Whether they were intertidal or subtidal or found on the fringing mangroves, these animals were an integral part of the aquatic ecologies of New South Wales' estuaries. Oysters are only one of the animals that benefited from the fine balance overseen by Aboriginal people all along the coastline. Extensive middens are the material testimony to the temporal and spatial scale of life in New South Wales' estuaries, both before and after the arrival of settler colonial life and politics.[3] Over the second half of the nineteenth century, these estuaries and their people saw rapid, destructive, and ultimately permanent damage under new forms of economic organization imposed by settlers. Ecological dynamism, always a feature of these places, was overlaid with settler dynamism, a different thing altogether.

In the second half of the nineteenth century, Eastern Australian estuaries supported a thriving oyster industry. This industry experienced three chronological phases, each one emerging from problems created by using the estuaries for their economic productivity rather than as interconnected hubs of more-than-human life. In the midst of these changes, a mudworm was imported from New Zealand, and it is this creature that captured the imagination of twentieth-century scientists attempting to unravel the decline in oyster production in the 1880s.[4] Mudworms were articulated as an invasive species, a perspective that enabled humans to set themselves apart from the complex environmental problems associated with oysters in this period in settler life. Here, I take an alternative approach to this environmental history, by using the motif of adaptation rather than invasion. In doing so, I tease out the ecological networks that formed around oysters in coastal Australia. Adaptation provides a way to expand the historical narrative to incorporate more agents, along with a deeper understanding of the ways that estuaries changed, and, the ecological network responded.

The first Intergovernmental Panel on Climate Change report in 1992 established the twinned pathways of mitigation and adaptation to regulate expected anthropocentric responses to climate change.[5] Political scientists John Dryzek, Richard Norgaard, and David Schlosberg called for research efforts to move from the emphasis on mitigation to unraveling the challenges of adaptation in the present and future. They identify resilience as a core feature of adaptation in socioecological systems, defining it as "flexibility, creative adaptation and constructive relationships with ecological processes."[6] Libby Robin argues that one of the ways that historians can contribute to debates about the Anthropocene is through "precise history" of the interactions between humans and nonhumans. She goes on to say that these will help "to provide a context for responding to the planetary crisis of our times."[7] I answer this call with historical research into the ways that settlers enacted adaptation in times of environmental change.

I offer the case study of the nineteenth-century oyster fisheries of eastern Australian estuaries to show the ways that adaptation links environmental degradation to colonial capitalism. I take issue with the idea that adaptation might provide a heuristic escape clause from the low levels of uptake of mitigation strategies pointed out by Dryzek, Norgaard, and Schlosberg. In the case of oysters in the estuaries of New South Wales, adapting to change did not mean an improvement in the state of these ecosystems. The overarching goal of the adaptation was to continue business as usual, leading to the loss of the socioecological integrity of these aquatic zones. Where capitalism drove adaptation, change contributed to the slow violence of the Anthropocene. We must acknowledge these histories of exhaustive adaptation to apply a precautionary principle to climate change research.

Colonial capitalism in the "new world" places of Australia and New Zealand approached environments as inexhaustibly abundant. The relationship between capitalism, colonialism, and environment has always been a fraught one. As ecofeminist Vandana Shiva stresses, a stolen harvest resulted when the global market was prioritized over local systems of food production and environmental stewardship.[8] In the estuaries of New South Wales, this was precisely the ideology followed by settlers. Viewing estuaries through an economic lens, nature reduced to resources, was a particularly rapacious way of seeing a place stolen from traditional owners through a war over land and sea. The displacement of traditional custodianship to enable unfettered access to resources resulted in enormous upheavals and changes to all the environments in Australia. Whether settlers sold oysters for food or as a lime additive for mortar in building, they were commodities for market.

This chapter uses the story of the oyster industry and its adaptation to environmental change to argue that capitalism combined with a narrow vision of resource extraction in estuaries resulted in the recurring and intensifying depletion of the host environments. To understand these changes, I will examine the way dynamic ecological networks responded to changes in the aquatic ecologies of the New South Wales coast. A key factor in the capacity of settlers to constantly adjust and intensify their destructive regimes emerges by focusing on the new elements introduced as oysters progressively disappeared. Each of these introductions was equally an adaptation to environmental change.

Two things about oysters: First, they are powerhouse filters of estuary waters, each one pumping up to five liters per hour through their gills to sieve out tiny nutrients for food.[9] Oysters, then, are keystone species who, like their mangrove counterparts, contribute to water clarity and quality.[10] Estuaries dense with oysters and oyster banks created water columns that were as clear as they were deep.[11] They complement the action of the mangroves in filtering out the detritus as it enters the main channels of the river or flows out into the bays.[12] Together these animals and plants help to construct, create, and maintain animal spaces for the tiniest infusoria up to the largest predatory fish and seaborne mammals like dugong and dolphins.[13] Second, oysters provide habitat and food for other animals of the estuary. The oyster reefs are shelter for fish of all sizes and other mollusks and shellfish. Many of these animals evolved skills enabling them to open the shells to include oysters in their diets. In 1892, the *Sydney Illustrated News* included images of eighteen species called "enemies" of the oyster. While it is true that these animals all predate on oysters, this illustration also gives us a glimpse into the rich array of fish and shellfish present in estuaries. Oysters interacted with the tiny mudworms and small whelks, as well as larger animals such as fish like leather jackets and mud crabs.[14] Furthermore, reef structures in water channels, creeks, and tributaries also changed the water flow, slowing water as it reached the estuary.[15] Such features made oysters an abundant food for Aboriginal people; one that was threatened by the arrival of settlers.

Estuaries as Contact Zones

Settlers very sparsely populated the New South Wales coast over the nineteenth century. Fishing industries were artisanal and low-scale ventures worked by small populations in comparison to inland migration and settlement connected to pastoralism and agriculture.[16] Nevertheless, the arrival of fishing fleets created contact zones between Aboriginal peoples and settlers. A contact zone is normally thought to be the arrival of the British in 1788. However, Heather Goodall's work on the European invasion of Aboriginal Australia shows the problems associated with the historical periodizing of contact violence.[17] Instead of fixing the moment of the arrival of the First Fleet in 1788 as a metonym for all contact, she argues for a more nuanced understanding of the experience of invasion.[18] In fact, the frontier reached different places at varying temporal and spatial scales, mapping onto the different kinds of economic ventures in specific ways for particular places.[19] Penny Edmonds extends this to the urban spaces of Melbourne, showing the spatial variability that created pockets of race violence depending on the value of places in the colonial economy.[20] It took some time for estuaries to be valued in the colonial economy. Estuaries, regarded by many as swamps, were not a coveted type of land for settlers, who preferred land that could be farmed, pastured, or cleared (as was the case in forestry). As Paul Shepard so eloquently expresses, swamps have a long history in Western cultures as full of "the fearful potential of organic fructification and the muckiness of wet decay." Shepard reveals the binary thinking of settlers where swamps were "opposed to the order of the farm or the city itself."[21]

Aboriginal people understood estuaries in entirely different ways. With honed traditional ecological knowledge, they could negotiate the water and land variability. They knew how to access the rich and varied food sources. They could establish large camps in culturally and physically appropriate places within estuaries and knew the time-honored pathways and songlines that led in, through, and out of the mangroves, saltmarsh, and wallum forests. Such maps held within the oral traditions of estuary life established tribal boundaries, including delineating special gendered areas (such as birthing or funerary places). Aboriginal people continued to read estuaries for these laws through the nineteenth century.[22] As settlers consumed other spaces for economic development, estuaries took on new meaning in the contact zones for Aboriginal people: they became sanctuaries. The wetlands and estuaries were refuges precisely because settlers saw them as wastelands. Godwin argues that such holes in the colonial maps enabled Aboriginal people to stay within the boundaries of traditional lands.[23]

European settlers also saw rivers as important places. Each New South Wales river supported the expansion of resource economics, as migrants moved out of urban centers like Sydney and Newcastle. Settlers used rivers as transport routes for commerce of all kinds. Colonial surveyors located townships upstream of the estuary, where the brackish water changed to consistent fresh, assisting the viability of settlements. However, Ray Kerkhove shows how settlers co-located the towns in southeast Queensland with the large Aboriginal camps adjacent to sources of fresh water.[24] The location of townships also demonstrates the spatial preference for pastoral and agricultural activities that relied on solid land.

Estuaries were considered nonproductive when compared to other kinds of land settlement and settlers largely left them alone. Heather Goodall and Alison Cadzow have shown in the Georges River catchment how the early maps of the area mark out the roads, farms, and towns, leaving the difficult terrain as untenable spaces, seemingly empty. Goodall and Cadzow use the records and oral testimony of Aboriginal people of the Georges River to show how this mapped emptiness was lively with Aboriginal life. Inhabitants included the people of that country, but also refugees from other places of the frontier war who knew these estuaries to be zones that white settlers did not understand. Swamps, sandstone escarpments, and estuary forests thus became places of survival.[25] Oystering from the 1850s represents the fluid frontier moving into the estuaries of New South Wales.

Harvesting Oysters

The first wave of fishing ventures in these estuaries relied on the simple collection of oysters from reefs and beds found in the zones between the river and the ocean. Six Sydney businessmen held licenses for the collection and sale of oysters in the capital. They paid laborers on a piecework basis for each bag containing 120 dozen oysters. These men and their families moved from estuary to estuary to work the oyster reefs. Oysterers collected from the mud flats and intertidal zones by hand at low tide and used tongs or dredges to access the permanently submerged reefs. These practices quickly depleted the available stock. Dredging destroyed the deep banks in as little as three years. Laborers rammed stakes into the reefs to break the compacted shells, allowing them to haul thousands of bags of oysters out of the estuaries.[26] Once complete, oysterers called this skinning a reef,[27] echoing the death of other indigenous animals of the sea country of the Aboriginal language groups of the coast. Up to forty men, women, and children worked each area barren, before moving on to the next estuary.[28]

The changes associated with oyster harvesting induced anxious concern for the viability of the oyster trade in New South Wales and with it, anxiety about the settlement of coastal places for human populations. One fix for this dual problem was to support new forms of oyster culture. Fisheries Inspector Alexander Black reported in 1877: "residents seem anxious for the introduction of culture; (they) . . . also express a desire that Government should provide sound and detailed information on culture."[29] Here, a conflict arose between the transient oysterers working for agents from Sydney and those who advocated that leases be made available for locals in local rivers. In responding to the 1877–78 Oyster Culture Commission, residents of Clarence, Shoalhaven, and Camden Haven Rivers petitioned for changes to the proposed new oyster fisheries legislation. Locals in these areas made a very clear distinction in favor of the creation of leases rather than continuing the arrangements that encouraged the "skinning of natural beds." They argued that the license system created a "hulking thriftless class of men, who would prowl about the fisheries to plunder the cultured beds."[30] Instead, this group promoted the introduction of leases for fixed areas along the foreshore. Residents made a direct comparison to "Whitstable oyster fishery and other oyster, pilchard and herring fisheries in Britain."[31] A change to leasehold arrangements, they claimed, would contribute to building fishing communities into towns as well as providing on-the-ground stewardship of oyster culture.

The movement of ideas and understandings of oyster culture drew in a myriad of different types of knowledge. This included scientific explorations of how to improve oyster growth through industrial applications. In one sense, this fits Roy MacLeod's formulation of the moving metropolis where the sciences "reflect and mediate the changing perceptions of vested interests" of capitalism to create a moving metropolis.[32] However, this narrow vision focuses on the politico-economic forces and ignores the material effect of these vested interests in the river systems. The scientists involved in this work were operating from local institutions to assist in the resolution of problems at a local—not global—scale. It is important to note that science was only one player in a broader network. As we shall see in the following sections, these changes in oystering and oyster culture also drew in nonscientists, especially those with Chinese and Aboriginal knowledge, and these local networks created the specificity of *each* estuary where oysters were grown for profit.

The Beginnings of Cultivation

A shift from licenses held by dealers in Sydney with dredgers working on a piecemeal basis to the introduction of surveyed and marked areas of oystering came with the *Fisheries Act 1881*. Oyster culture, along with marked-out leasehold areas fixed settlers in place, giving them settlers' rights to this stolen sea country. The language of improvement used in association with oyster culture by governments and leaseholders echoes the land-based activities of settler Australia.[33] A universal lease condition for rural Australia was the requirement to improve the land. As a rule, this related to clearing the land of trees, keeping weeds under control, building fences, or introducing fodder species for livestock. The notion of cultivation has deep ties with agriculture in the way it was used to justify encroachment into contact zones that crossover in these new relations with estuaries and oysters.[34]

Oyster leases, in replicating land tenure, also replicate broader ideas about the settlement of Australia as a method of dispossession, here expanding to encompass estuaries and coastal places. Those petitioning for such changes linked this settling process with "encouragement to a respectable and industrious class of men to engage in that pursuit."[35] Settlers applied the same ideas of converting worthless land to industrious estates, using the labor of the lessee to generate this transformation. Labor plus land created a kind of respectability tied to the colonial project of settlement and dispossession in its broadest political sense. In the estuaries, leases came in two forms. The first was the dredge sections that consisted of a larger plot, usually in the deeper channels of the estuary. Second, oyster leases were marked out in uniform plots usually closer to the shoreline where cultivation could take place. It is this second form that became associated with oyster culture.

The first attempts at cultivating oyster growth focused on the areas already depleted. Once dredging the reefs destroyed the oyster density, oysterers scattered brood stock over the skinned areas, to little effect. In Port Macquarie, Andrew Barber, a veteran dredger of thirty years, found that when he returned to a skinned reef after three years, oysters had not recovered.[36] Initially, oyster culture consisted of moving smaller oysters from deep water to higher ground to monitor growth. Many oyster farmers also gathered oysters from mangrove pneumatophores and placed them in the deeper water. These attempts at translocation represent deliberate ecological change as part of settlement, a change from simple collection to the stewardship of overseeing the fattening of oysters for market. Translocation was an older technology borrowed from Europe, where Pliny first described artificial culture of this kind in 95 B.C.[37] In addition, immature oysters were moved from one estuary to another, generally from north to south within Australia. Oysterers also looked to New Zealand, a nearby white settler colony, where the same oyster species grew, to supply this cultivation stock. In New South Wales, it was cheaper to translocate from New Zealand than to rely on Australian stock.[38] Settlers presumed that these transfers would have the same impact as those already being moved within the Australian ecological networks—that is, it would present no problem at all.

The Mudworm Invasion?

Intercolonial exchanges between Australia and its near neighbors represent a vital concept of community in a time of environmental damage. As a result of their common colonization, Australia and New Zealand shared a range of different kinds of knowledge, science, and governance. Biological exchanges in the form of both botanical and zoological specimens also slotted neatly into broader networks of empire. Each settler state was adapting to the changes that extraction economies experienced in the early part of the nineteenth century. Neither colony recognized the specificity of estuarine ecologies as settlers treated oysters as the "same" between the two places. Australians introduced oysters from New Zealand into twelve estuaries along the New South Wales Coast between 1880 and 1889.[39] They overlooked the presence of mudworm in the New Zealand oysters in this intercolonial exchange. Damien Ogburn, Ian White, and Daryl McPhee demonstrate the correlation between introductions from New Zealand and the outbreak of "mudworm disease." They show a rapid decline in volume of oysters from these places over the decade of the 1880s. From these twelve rivers, they argue, mudworms spread to other estuaries.[40]

Mudworms cohabitate with oysters all over the world, including Australia and New Zealand. Larvae of the *Polydora* worm enter the oyster when it opens its valves to feed. There they settle in the corner of the shell far enough from the animal to escape initial detection. They are not interested in the oyster per se, but in the oyster shell as an estuary dwelling. Mudworms build tiny tubes and coopt the current generated as the oyster feeds to draw food for itself. Once the oyster *does* detect the worm, a furious maelstrom of energy exchanges between the two. The oyster grows shelly matter over the worm, and the worm fights back by relocating above the new shelly matter, creating a honeycomb within the shell. Some oysters became exhausted in this fight. When these shells were shucked open, the pungent rot of mud and dwindling body of the oyster greeted the consumer. Because only a small proportion of oysters were made inedible by mudworm, no action was required. That was until the New Zealand oysters arrived with their companion travelers, *Polydora* worms. After the New Zealand oysters arrived, there is evidence of increased incidence of mudworm in harvested oysters. This in turn amplified the level of concern among settlers.

It is true that mudworms from New Zealand entered an already damaged aquatic ecology. In addition to the direct loss of oyster banks from dredging, land use practices like forestry, large-scale vegetation clearance, waste from gold fossicking,[41] and the introduction of domestic hooved animals created new kinds of sediment loads. Heavier loads of run-off carried through the estuaries during frequent summer storms and regular flood events.[42] While some fish and shellfish populations, including oysters, declined because of these changed conditions, mudworm thrived. These combined changes were particularly devastating to the larger subtidal flat oysters *Ostrea angasi*. The depletion of oysters and the changed environmental conditions, including a higher population of mudworms, resulted in fewer oysters for trade in the urban marketplace. The viability of the trade was under threat, requiring an adaptation to these new agents in estuary life. Mudworms came into the spotlight as an environmental scapegoat.

In 1890, Australian Museum zoologist Thomas Whitelegge undertook the first detailed examination of both mudworm and its impact on oyster trade under local Australian conditions.[43] Whitelegge's report was drawn from two weeks' fieldwork in the oyster estuaries around Newcastle, north of Sydney. There he spent time talking to local oysterers, the Gibbons[44] and Anderson families, while traveling with the local inspector of fisheries, Henry Curan. He immersed himself in these estuary places, watching for wind, tides, animals, and people. He collected mudworms, their eggs, larvae, and adult forms, as well as a variety of infected oysters. He recorded the details of the mudworm life cycle. He detailed how the mudworm set up house in the oyster's shell, how the oyster responded, and what happened as the mudworm grew. It was clear from his observations that while the relations between mudworm and oysters were evident in estuary life all over the world, for some reason it was out of kilter in New South Wales. Mudworms were especially prevalent in the oysters located on the mud flats "about (and below) the low-water mark."[45]

Whitelegge then set about formulating and carrying out some experiments to test the limits of mudworm life under a range of different conditions. One of the variables was salinity, where he found that mudworms could not tolerate too much fresh water. A second set of experiments replicated tidal conditions, because Whitelegge had observed more mudworms in permanently submerged oysters. Here he found that he could harness the tides to create favorable conditions for the oyster. Mudworms spurned those oysters that spent part of their day dry. His 1890 report was originally delivered to the Commissioners of Fisheries and reprinted in the Australian Museum records the same year. Its reprint gave it a far wider readership.[46] Additionally, newspapers printed extracts of these findings.[47] As Whitelegge had worked closely with oysterers to produce this knowledge, it was circulated back to these same leaseholders and inspectors through this print media.

Technologies of Oyster Culture

Paralleling the varying temporal and spatial scales of invasion, the technological adaptation of culture occurred at erratic times and in different places. At this time in New South Wales, the case for increasing the technological aspects of oyster culture was gaining traction in fisheries management. In 1876–77, a Royal Commission into Oyster Culture tapped into the international research into cultivation. The Royal Commission gathered reports from fisheries scientists in Britain, France, and America to formulate a plan for reinvigorating overworked reefs.[48] Led by Thomas Holt, the commission focused on two aspects of international oystering: the introduction of cultch and the change from licenses held in the city to leases managed in place. Cultch is any rough matter to which swimming oyster larvae attach. Advice from international advocates of oyster culture was to introduce artificial cultch into estuary systems as a way of controlling the biology of the oyster. Oyster larvae swim free for three or more weeks, before settling and attaching to aquatic substrate. As they attach, they transform from larvae to spat. Oysterers included throwing oyster shells back into the water, to placing rocks, slabs, or timber into the intertidal area to increase the substrate of the estuary.

Using cultch to enhance populations of oysters predates nineteenth-century science. Romans, great lovers of oysters, used this method to ensure plenty of oysters for highbrow feasts. In the mid-nineteenth century, however, scientific interventions into oyster culture began in France. To increase production, scientists focused on systemization of the processes involved in oystering.[49] In 1851, the French professor of embryology Victor Coste, under orders from Napoleon III, surveyed the oyster-producing regions along the French and Italian coastline. He visited Lake Fusaro in Italy, where the oyster culture was thought to trace a line of inheritance directly to the Romans.[50] Coste proposed a system for France to "restock its ruined beds, to revive those which (were) extinguished, to extend those which prosper, (and) to create new ones wherever the nature of the bottom will permit their establishment."[51] He then set about systematically using cultch in conjunction with enhancing the established system of claires or fattening ponds. In this way, the French system controlled the spat fall and provided stewardship for the oysters.[52] These systems revived the oyster industry and ensured regular supply, especially for the capital in Paris.

At the same time Coste made these recommendations in France, Thomas Eyton described changes to the oyster industry along the coasts of England, Ireland, Scotland, the Isle of Man, and Wales.[53] In 1858, Eyton also argued for the introduction of cultch when preparing new oyster beds or renovating those that were "exhausted."[54] Thomas Huxley, commissioner of sea fisheries, came to this same conclusion about depletion, but was positive that modern oyster culture would prevail.[55] Along both coastlines, the introduction of cultch provided a substrate in the subtidal zone for the collection of the spat. The Royal Commission into Oyster Culture in Australia drew on this work from Europe. The final report handed to the New South Wales government in 1877 reproduced illustrations from the Irish Commission of Inquiry into the Method of Oyster Culture, including images of spat collectors called "fascines," from Lake Fusaro, Italy, and St. Brieux, France, as well as the claires of France and Ireland.[56] In addition, the Fisheries Commission printed key scientific papers on oyster culture, especially reproduction, which included Huxley's work, in addition to papers considering American scientific research. These sets of reprints were circulated for a limited audience of oyster leaseholders. While we cannot know how far this knowledge traveled, oysterer Henry Woodward used material gleaned from Huxley's paper to reinterpret the problem of oyster decline in 1883.[57]

On the other side of the Atlantic, fisheries scientists were also experimenting with systematizing oyster culture to increase the productivity of oyster reefs. In the 1880s, William K. Brooks joined the new Johns Hopkins University where he worked closely with the fisheries sector to match biological knowledge with economic questions. Brooks promoted the idea that "nature need not be exhausted," pointing to hatchery technologies as supplements to the reproduction of fish. He believed that these techniques were perfect for the oyster fisheries of America, especially in his local estuary, the large and complex Chesapeake Bay. According to Christine Keiner, Brooks believed "that the tools of reproductive biology could be used to decrease the reliance on the vagaries of nature."[58] One element of this theory included the speculation that oyster fields could expand into new parts of the bay largely through the systematic introduction of "cultch."[59] Additionally, Brooks was an advocate for privatizing the oyster fishery through lease arrangements. When fisheries inspector Alexander Black was reviewing oysters in the Richmond River, he compared them to those he had experienced in the oyster culture of Europe and America.[60] In this way, the technological developments of oyster culture in the northern hemisphere transferred to New South Wales. As Peter Minard argues in relation to salmonid acclimatization in Australia, settlers looked to both Europe *and* North America for guidance about enhancing the depleted environments of colonial Australia.[61]

The introduction of cultch was designed to intervene in the biological life cycle of the oyster by collecting the spat. Initially this was to increase the substrate, but as the technology changed it was specifically to collect the spat such that the oysterers could move it to other locations. Given that each oyster expels more than 100 million sperm or up to 20 million eggs into the water, spawning provides an important function in the food cycles of estuaries for local fish. Only a tiny proportion survives as spat. Neither science nor industry acknowledged this aspect of estuarine life as vital to the safeguarding of the fish communities in densely populated estuaries. John Hughes from Cape Hawke observed "numerous fish collected amongst it, seemingly to feed on it."[62] But the signal master from the same area declared that "he never knew a dredger who did" care about whether fish ate the spawn.[63] For those looking at spat as a resource, each young oyster lost into the waterways was potential income flowing out to sea. Capturing spat was one way to stabilize the trade in oysters. Along with the destruction of oyster reefs, however, removing larvae is clear evidence of the shortsightedness of the settlers. They were only able to see extraction and aquaculture to serve their immediate needs rather than the functioning of the ecology for the sustenance of the denser networks of humans and nonhumans.

Mudworm changed the biological circumstances in Australian estuaries so that a direct transfer of northern hemisphere techniques was not possible. It didn't matter whether oysterers used rocks, old oyster shells, or any other introduced material that replicated the bottom surfaces of the river. As the submerged oysters grew, the mudworm populations grew with them. Australian oysterers needed to do something different from their counterparts in the northern hemisphere. They had to adapt what they knew—about technology, and the life cycle of the oysters and their local environments—to create new methods of approaching the oyster culture. There was one technology used in Europe that provides a clue for what Australian oysterers did next. In Lake Fusaro of Italy and along the coast of France, oysterers used fascines or bundles of sticks gathered together and fastened to the sea or riverbed. They were used to great effect in low-flow water bodies like lakes.[64] This technology kept oysters underwater, and in Australia this made them vulnerable to mudworm. Local oysterers adapted the fascines design imported from Europe to the different conditions in New South Wales. Rather than held under water, the Australian fascines were laid in the intertidal area. In addition to fascines, there was a second method that Australian oysterers adapted, this time from another part of the world—China.

Oyster culture is recorded as present in Chinese estuaries for over 2,000 years.[65] Although nineteenth-century Fisheries Department records show the presence of Chinese workers in Australian estuaries, their influence in aquaculture is an underexplored feature of our environmental history.[66] In his research on Chinese fishing technologies in Australia, archaeologist Alister Bowen highlighted the sophisticated fish-drying industry in Victoria. He also speculated that the fishing industry of the southern New South Wales coast included Chinese workers.[67] In 1880, fishing merchant Chin Ateak told the Royal Commission into Fisheries of salting, pickling, and smoking fish in enterprises at Port Stephens, Jervis Bay, and Merimbula. He detailed large stations employing European fishers and Chinese curers between the 1850s and 1870s; however, it is unclear from his evidence whether he was involved in oystering.[68]

The English scientist William Saville-Kent surveyed the Australasian oyster industry between 1884 and 1892. In his lecture to the Australian Association for the Advancement of Science in Christchurch, New Zealand, in 1891, he pointed out that Australians could apply Chinese methods for cultivation and spat collection to the rock oyster. He described spat collection using bamboo sticks: "With oyster shells inserted into the cleft ends which are then stuck upright into the mud, the shells being raised a few inches above the surface. The spat soon collects on the shell, and when a sufficient size is removed and distributed over the fattening beds."[69] Along the Chinese coastline, bamboo tents were also set into the soft mud of the intertidal zone for spat collection.[70] Once spat had anchored, each bamboo stick was placed individually into the tidal zone.[71] Bamboo was a local, easily accessed, fast-growing, and renewable material in places like the Pearl River Delta.[72] These transnational influences, all from the northern hemisphere, came together in a new form of spat collection specific to the estuaries of New South Wales.

Oystering and Mangroves

Stick culture using mangrove timber became the new technology that adapted the Australian industry to the changed conditions. Where bamboo was the timber of choice in China, in Australia, the black mangrove *Aegiceras corniculatum* took its place as a durable local wood. These techniques were used in oyster culture in the Hawkesbury River, just north of Sydney, the Port Stephens area, and the Manning River in the northern districts of New South Wales. Here we see a combination of a whole range of transnational elements interwoven into the estuary itself. These spat collectors either echo the design of the fascines from Europe or were set into the mud for stability in the fashion of the Chinese tents. In both cases, these structures were solid in the changing intertidal zone; they kept the oysters dry for part of the day and thus made them uninhabitable for mudworms. They are also biomimicry, replicating the form and function of aerial mangrove roots on which spat collected under estuary conditions.

Once the spat were collected, whether on fascines or tents, the sticks were separated and relocated into the estuary's tidal flats as individual growing agents. Where bamboo was the timber of choice in China, in Australia mangroves took its place. Black mangroves were water-durable timber evolved to live in brackish water and their sticks survived in the water for around three years before they started to break down; fortunately, the same time required for oyster maturation. Oysterers took advantage of this biological decay. They collected each stick and hit it on the gunwale of the punt, causing the oysters to fall off. They then threw the sticks back into the estuary. Such collectors for spat used technologies that drew on a range of transnational and local elements that together enabled the continued farming of the foreshores. If the drop in oyster production was due to mudworm in the 1880s, as Ogburn, White, and McPhee show, then the recovery of the industry in the 1900s was aided by stick culture.

There are forty-one species of mangrove in Australia.[73] Along the coastline where Sydney Rock Oysters occur, two species dominate: the grey mangrove *Avicennia marina*, which is a large tree; and black mangrove *Aegiceras corniculatum*, a smaller, generally 2- to 3-meter-tall shrub that forms dense thickets. In oyster spawning season, usually spring and summer months, spat settle on mangroves in large numbers. Because of their position in the tidal zone, the aerial peg systems of grey mangroves and the bark of black mangroves are abundant settlement areas for oyster spat. Once oysterers identified this potential, black mangroves became the main variety harvested for use as oyster sticks as they found that grey mangroves did not last until the oysters matured.[74] Grey mangroves were susceptible to quick consumption by an estuarine woodworm known as cobra. Black mangroves can occur as "pure dense stands" both along riverbank and also landward of more mixed mangrove stands. These dense stands on the edge of estuaries made them easier to acquire.[75] Harvesters cut the black mangroves down to uniform four- to six-foot-long sticks for use in fascines and tent formations. After six months, oysterers moved the securely attached spat into the more brackish water and staked individual sticks out in dense runs.

Patrick Dwyer from the New South Wales Department of Primary Industries has compiled a thorough analysis of the impact of removing black mangroves for oyster culture between 1900 and 1950.[76] He points out that the *Fisheries Act 1902* incorporated the laying down of cultch into the lease arrangements. The introductory remarks of the Fisheries and Oyster Farmers of New South Wales Report of Commissioners for 1903 noted that: "up to the passing of the present Act, there was no power to compel lessees to improve their leased areas. . . . Now, however, it is compulsory for lessees to cultivate their leased areas within twelve months of issue, lay down suitable material for the collection of spat, and failing to do this, the leases will be liable to cancelation."[77]

Although suitable materials included stone, shells, shingles, and tiles, by the 1900s black mangroves were the preferred cultch for spat collection. This method was "strongly recommended" by the government fisheries inspector in his report on the Richmond River in 1909.[78] Initially, mangroves were harvested locally to supply sticks directly to oyster leases in situ. In a striking replication of the brood stock of oysters moved from estuary to estuary, the mangroves suffered the same fate. By the 1920s, an export trade grew to supply mangrove sticks to other locations in New South Wales. Dwyer estimates, for example, that 3 million black mangrove sticks were exported from the Richmond River alone, desecrating up to 20 kilometers of river length. Dwyer found that oysterers were forced to stop using this timber around 1950, because they had not been mindful of replenishing tree stock, only stripping it out for their immediate use.[79] They had replicated the decimation of oyster banks by destroying mangrove forest, thus compounding the ruin of these aquatic ecologies.[80]

Conclusion

Adaptation is a critically important motif to help us to think through human-induced environmental change in estuaries. A point of adaptation, here the response to the invasive mudworms, is also a means to unravel an ecological network. We articulate these changes as a crisis of the economy when, in fact, at each stage settlers multiplied the effects of environmental change across species and places. Each failed environment was a result of settlers' management of the solution to the previous problem through knowledge limited by colonial capitalism. This reconsideration of the nineteenth-century oyster industry in Australia shows oysters to be agents through which we can see the anxious adaptations of colonization. Settlers deployed a recurrent, but always shifting mode of extraction as their presence intensified and precipitated environmental conditions. The methods and materiality of commercial oyster culture allows us to trace the intensifying enmeshment of people and place within the processes of settlement. Settlers failed to acknowledge the estuary entanglements, both temporal and spatial. It did not occur to them to replenish what they took. Instead, they persisted with a confident expectation of serial adaptations without acknowledging the greater environmental problem of colonization for profit in a place stolen and different.

Notes

1. Peter S. Roy, Robert J. Willliams, Alan R. Jones, Iradj Yassani, Philip J. Gibbs, B. Coates, Ronald J. West, Peter R. Scanes, J. P. Hudson, and Scott Nichol, "Structure and Function of South-East Australian Estuaries," *Estuarine, Coastal and Shelf Science* 53 (2001): 360.

2. Eric Wolanski, ed., *Estuaries of Australia in 2050 and Beyond* (Dordrecht, Netherlands: Springer Science+Business Media, 2014); Lynn Turner, Deiter Tracey, Jan Tilden, and William C. Dennison, *Where River Meets Sea: Exploring Australia's Estuaries* (Collingwood, VIC: CSIRO Publishing, 2004); John W. Day Jr., Byron C. Crump, W. Michael Kemp, and Alejandro Yáñez-Arancibia, eds., *Estuarine Ecology* (Hoboken, NJ: John Wiley & Sons, 2012).

3. Ian Hoskins, *Coast: A History of the New South Wales Edge* (Sydney, NSW: NewSouth Publishing, 2013), 51–54.

4. Damian M. Ogburn, Ian White, and Daryl P. McPhee, "The Disappearance of Oyster Reefs from Eastern Australian Estuaries, Impact of Colonial Settlement or Mudworm Invasion?" *Coastal Management* 35 (2007): 271–87; Michael W. Beck,

Matthew C. Kay, Hunter S. Lenihan, Mark W. Luckenbach, Caitlyn L. Toropova, Guofan Zhang, Ximing Guo, Robert D. Brumbaugh, Laura Airoldi, Alvar Carranza, Loren D. Coen, Christine Crawford, Omar Defeo, Graham J. Edgar, and Hancock Boze, "Oyster Reefs at Risk and Recommendations for Conservation, Restoration, and Management," *BioScience* 61 (2011): 107–16; John A. Nell, "The History of Oyster Farming in Australia," *Marine Fisheries Review* 63 (2001): 14–25.

5. In the report, the word "limitation" is used instead of mitigation. "Mitigation" later became the word taken up in policy and politics across the globe. Adaptation has remained a keyword in policy formulation: International Panel on Climate Change, *Climate Change: The IPCC Response Strategies*, edited by IPCC Working Group 3 (New York: World Meteorological Organization and United Nations Environment Program, 1990), xxvii.

6. John Dryzek, Richard B. Norgaard, and David Schlosberg, *Climate-Challenged Society* (New York: Oxford University Press, 2013), 131.

7. Libby Robin, "Histories for Changing Times: Entering the Anthropocene?," *Australian Historical Studies* 44 (2013): 335.

8. Vandana Shiva, *Stolen Harvest: The Hijacking of the Global Food Supply* (Cambridge, MA: South End Press, 2000), 37–56.

9. Beck et al., "Oyster Reefs at Risk."

10. Jennifer L. Ruesink, Hunter S. Lenihan, Alan C. Trimble, Kimberly W Heiman, Fiorenza Micheli, James E. Byers, and Matthew C. Kay, "Introduction of Non-Native Oysters: Ecosystem Effects and Restoration Implications," *Annual Review of Ecology, Evolution, and Systematics* 36 (2005): 646–48.

11. Wayne O'Connor, principal research scientist, aquaculture research, Port Stephens Fisheries Insittute, personal communication, 14 October 2015: in biological terms, one of the major roles of oysters is benthic pelagic coupling. It is implicit here but understated. While oysters do filter out particles on which to feed, they essentially do so selectively; that is, only a portion of what is filtered is consumed. The remainder, and it can be a large portion, of what is filtered is rejected from the shell as "pseudofeces" (aggregates of the suspended particles that were in the water column). Given the numbers of oysters that did exist and their filtering capacity, this is a mechanism for large-scale removal of material from the water column to the benthos where it provided nutrients/food for many other species.

12. Donald S. McClusky and Michael Elliot, "Primary Consumers: Herbivores and Detrivores," *The Estuarine Ecosystem: Ecology, Threats, and Management* (Oxford: Oxford University Press, 2004): 53–72.

13. William Saville-Kent, *Oysters and Oyster Fisheries of Queensland* (Brisbane: Queensland Government Votes and Proceedings, 1891), 605–7.

14. "Group of Oysters and Their Enemies," *Sydney Illustrated News*, 3 September 1892, 7.

15. Roger I. E. Newell and Evamaria W. Koch, "Modeling Seagrass Density and Distribution in Response to Changes in Turbidity Stemming from Bivalve Filtration and Seagrass Sediment Stabilization," *Estuaries* 27 (2004): 793–806.

16. Hoskins, *Coast*, 149–82.

17. Heather Goodall, *Invasion to Embassy: Land in Aboriginal Politics in New South Wales, 1770–1972* (St. Leonards, NSW: Allen & Unwin in association with Black Books, 1996).

18. Gammage makes this epistemic move in Bill Gammage, *The Biggest Estate on Earth: How Aborigines Made Australia* (Sydney, NSW: Allen & Unwin, 2011).

19. Luke Godwin, "The Fluid Frontier: Central Queensland 1845–63," in *Colonial Frontiers: Indigenous-European Encounters in Settler Societies*, edited by Lynette Russell (Manchester, UK: Manchester University Press, 2001), 101–18; Jessie Mitchell, "'Great Difficulty in Knowing Where the Frontier Ceases': Violence, Governance, and the Spectre of India in Early Queensland," *Journal of Colonial History* 15 (2013): 43–62.

20. Penelope Edmonds, "The Intimate, Urbanising Frontier: Native Camps and Settler Colonialism's Violent Array of Spaces around Early Melbourne," in *Making Settler Colonial Space: Perspectives on Race, Place and Identity*, edited by Tracey Banivanua-Mar and Penelope Edmonds (London: Palgrave Macmillan, 2010), 129–54.

21. Paul Shepard in Geoff Park, "'Swamps Which Might Doubtless Easily Be Drained': Swamp Drainage and Its Impact on the Indigenous," in *Making a New Land: Environmental Histories of New Zealand*, edited by Eric Pawson and Tom Brooking (Dunedin, New Zealand: Otago University Press, 2013), 177; see also Rodney James Giblett, *Postmodern Wetlands: Culture, History, Ecology* (Edinburgh: Edinburgh University Press, 1996); Barbara Hurd, *Stirring the Mud: On Swamps, Bogs, and Human Imagination* (Boston, MA: Beacon Press, 2001).

22. Margaret Somerville and Tony Perkins, *Singing the Coast* (Canberra, ACT: Aboriginal Studies Press, 2011); Robin A. Wells, *In the Tracks of the Rainbow: Indigenous Culture and Legends of the Sunshine Coast* (Sunshine Beach, QLD: Gullirae Books, 2003).

23. Godwin, "The Fluid Frontier," 107–8.

24. Ray Kerkhove, *Aboriginal Camps of Greater Brisbane: An Historical Guide* (Brisbane, QLD: Boolarong Press, 2015).

25. Heather Goodall and Allison Cadzow, *Rivers and Resilience: Aboriginal People on Sydney's Georges River* (Sydney, NSW: UNSW Press, 2009).

26. Alexander Black, "Oyster Culture Commission," in *Oyster Culture Commission, Report of the Royal Commission Appointed to Inquire into the Best Mode of Cultivating the Oyster, of Utilizing, Improving, and Maintaining the Natural Oyster Beds of the Colony, and Also as to the Legislation Necessary to Carry Out These Objects*, edited by Thomas Holt (Sydney: New South Wales Legislative Council, 1886–7), 54.

27. Langham, "Oyster Culture Commission," 69–78.

28. Black, "Oyster Culture Commission," 53–69.

29. Black, "Oyster Culture Commission," 56–57.

30. Oyster Fisheries, *Petitions from Residents of Shoalhaven, Clarence and Camden Haven Fishery Districts* (Sydney: New South Wales Votes & Proceedings, Legislative Assembly, 1877–8), 880, 882, 884.

31. Oyster Fisheries, *Petitions from Residents*, 880, 882, 884.

32. Roy MacLeod, "On Visiting the Moving Metropolis: Reflections on the Architecture of Imperial Science," *Historic Records of Australian Science* 5 (1982): 14.

33. Dennis N. Jeans, *An Historical Geography of New South Wales to 1901* (Sydney, NSW: Reed, 1972).

34. Cameron Muir, *The Broken Promise of Agricultural Progess: An Environmental History* (London: Routledge-Earthscan, 2014).

35. Oyster Fisheries, *Petitions from Residents*.

36. Black and Langham, "Oyster Culture Commission," 59.

37. Rebecca Stott, *Oyster* (London: Reaktion Books, 2004), 24.

38. John A. Nell, "The History of Oyster Farming in Australia," *Marine Fisheries Review* 63 (2001): 14–25.

39. Ogburn, White, and McPhee, "The Disappearance of Oyster Reefs."

40. Ogburn, White, and McPhee, "The Disappearance of Oyster Reefs."

41. Only the Moruya River on New South Wales Coast attracted gold miners. The bulk of the gold rushes were located beyond the Great Dividing Range in New South Wales and in Victoria and Queensland.

42. William Saville-Kent, "Oysters and Oyster Culture in Australasia," *Australasian Association for the Advancement of Science* (Christchurch, New Zealand: 1891), 8.

43. Thomas Whitelegge, "Report on the Worm Disease Affecting the Oysters on the Coast of New South Wales," *Records of the Australian Museum* 1 (1890): 41–54.

44. Gibbons had leases in New Zealand and is believed to be the likely vector for introduction according to Ogburn.

45. Whitelegge, "Report on the Worm Disease," 41.

46. Circulation of *Records of the Australian Museum*—4,000—population of NSW in 1890 was 1,113,275 (ABS). Aboriginal peoples were not counted in this figure. http://www.abs.gov.au/AUSSTATS/abs@.nsf/DetailsPage/3105.0.65.0012006?OpenDocument (9 June 2015).

47. The Oyster Disease: Caution to Oyster Leases," *Clarence and Richmond Examiner*, 4 March 1890, 4; "The Oyster Disease," *The Richmond River Herald and Northern Districts Advertiser*, 7 February 1890, 3; "The Oyster Disease," *Northern Star*, 1 March 1890, 2.

48. Christine Keiner, *The Oyster Question: Scientists, Watermen, and the Maryland Cheasapeake Bay since 1880* (Athens: University of Georgia Press, 2010); Thomas H. Huxley, "Oysters and the Oyster Question," in *Articles on the Propagation of Oysters Printed by Order of the Commissioners of Fisheries* (Sydney, NSW: Thomas Richards, Government Printer, 1883); W. Anderson, "Oyster Cultivation in Scotland," in *Fish and Fisheries: A Selection of the Prize Essays of the International Fisheries Exhibition, Edinburgh*, edited by David Herbert (Edinburgh: William Blackwood and Sons, 1882), 28–36; "Various Methods of Oyster Culture," in *Fish and Fisheries: A Selection of the Prize Essays of the International Fisheries Exhibition, Edinburgh*, edited by David Herbert (Edinburgh: William Blackwood and Sons, 1882), 17–27; Thomas C. Eyton, *A History of the Oyster and the Oyster Fisheries* (London: John Van Voorst, 1858).

49. Darin Kinsey, "'Seeding the Water as the Earth': The Epicenter and Peripheries of a Western Aquacultural Revolution," *Environmental History* 11, 3 (2006): 527–66.

50. Eyton, *A History of the Oyster and the Oyster Fisheries*, 251–52.

51. Quoted in Eyton, *A History of the Oyster and the Oyster Fisheries*, 38.

52. Kinsey, "'Seeding the Water as the Earth.'"

53. Eyton, *A History of the Oyster and the Oyster Fisheries*.

54. Eyton, *A History of the Oyster and the Oyster Fisheries*, 33–34.

55. Huxley, "Oysters and the Oyster Question," 32.

56. Holt, *Oyster Culture Commission*, 17–21.

57. Henry Woodward, *Oyster Culture in New South Wales* (Sydney, NSW: John Woods and Co., 1887).

58. Keiner, *The Oyster Question*, 61.

59. Keiner, *The Oyster Question*, 61–102.

60. Black, "Oyster Culture Commission," 54.

61. Peter Minard, "Salmonid Acclimatisation in Colonial Victoria: Improvement, Restoration and Recreation 1858–1909," *Environment and History* 21 (2015): 177–99.

62. Black, "Oyster Culture Commission," 64.

63. Black, "Oyster Culture Commission," 65.

64. Olivier Levasseur and Darin Kinsey, "The Second Empire Legacy of the French 'Culture' of Oysters," *International Journal of Maritime History* 20 (2008): 253–68.

65. Stott, *Oyster*, 74–76; Ximing Guo, Susan E. Ford, and Fusui Zhang, "Molluscan Aquaculture in China," *Journal of Shellfish Research* 18 (1999): 19.

66. Kuanhong Min and Baotong Hu, "Chinese Embankment Fish Culture," in *Integrated Agriculture-Aquaculture: A Primer*, edited by Marie Sol Sadorrra-Colocado (Rome: Food and Agriculture Organization of the United Nations, 2001), 25–29.

67. Alister Bowen, "Material Evidence for Early Commercial Fishing Activities on the Far South Coast of New South Wales," *Australasian Historical Archaeology* 22 (2004): 79–89; Alister Bowen, *Archaeology of the Chinese Fishing Industry in Colonial Victoria* (Sydney, NSW: Sydney University Press, 2012).

68. Chin Ateak, Evidence presented on 4 February 1880, *Report of the Royal Commission Appointed on the 8th January, 1880, to Inquire into and Report on the Actual State and Prospect of the Fisheries of this Colony; Together with the Minutes of Evidence and Appendices* (Sydney: New South Wales Legislative Assembly, 1880).

69. Saville-Kent, "Oysters and Oyster Culture in Australasia."

70. Brian E. Spencer, *Molluscan Shellfish Farming* (Oxford: Blackwell Publishing, 2002), 123–46.

71. Yingya Cai and Xuanhai Li, "Oyster Culture in the People's Republic of China," *World Aquaculture* 21(1990): 67–72; George C. Matthiessen, "Oyster Culture in the Far East," in *Oyster Culture*, edited by George C. Matthiessen (Oxford: Fishing New Books, 2001), 35–46.

72. The largest influx of Chinese migrants arrived from this region to Australia. I have found no documentary evidence to link an individual to the transfer of these methods from China to New South Wales. On one hand, the acknowledgment of Chinese presence in the fisheries reports is suggestive of transfer of knowledge about Chinese oyster culture. On the other hand, it is also possible that a European traveling between China and New South Wales facilitated the use of tents as spat collectors.

73. Mangrove Forest, Australian Forest Profiles, Australian Government, Department of Agriculture, www.agriculture.gov.au/abares/forestaustralia/profiles/mangrove-forest (2 June 2015).

74. "Mangroves—Oyster Culture, Little Known Richmond Industry, Big Shipments Made to Port Stephens," *Northern Star* 2 (March 1934): 6.

75. Red Mangrove, Common Mangroves, Queensland Government, Department of Agriculture and Fisheries, www.daf.qld.gov.au/fisheries/habitats/marine-plants-including-mangroves/common mangroves/river-mangrove (2 June 2015).

76. Pat Dwyer, "'Exhausted the Home Supply': Historical Harvesting of River Mangroves," Conference presentation, Consider the Oyster . . . again (Sydney, NSW: Sydney Environment Institute, University of Sydney, 2013).

77. Cited in Dwyer, "'Exhausted the Home Supply,'" 4.

78. "'Exhausted the Home Supply,'" 4.

79. Dwyer, "'Exhausted the Home Supply,'" 8.

80. By the time that ecologists started doing vegetation surveys in the 1970s, they were not even looking for these trees as part of the aquatic ecology. But they were not looking for the impact of oyster bank removal either. Dwyer is advocating for a reassessment of the baseline used for restoration of wetlands; I would argue that the destruction of oyster banks and reefs both need to be accounted for in this process.

Brumbies (*Equus ferus caballus*) as Colonizers of the Esperance Mallee-Recherche Bioregion in Western Australia

NICOLE Y. CHALMER

"Brumby" is the Australian term for feral wild horses (*Equus ferus caballus*). This distinctively Australian vernacular was first published in the *Australasian* (Melbourne) in 1880 as referring to wild horses in Queensland. Though the origin is obscure, it may have originated from the Pitjara Aboriginal language group in Queensland, whose word "baroombie" was reported as meaning unbroken or wild horse.[1] The ancestors of the Esperance Mallee-Recherche bioregion (EM-R) brumbies were brought there by the first Anglo settlers in the 1860s. Horses aided Anglo occupation and reengineering of Aboriginal and marsupial landscapes during colonialism by acting as part of a broader suite of biotic introductions that Crosby described as Europeans' "portmanteau biota."[2]

The landscapes colonized by horses and the Europeans who brought them were not wilderness in the common usage; they were humanized landscapes because they had been shaped for at least 45,000 to 50,000 years by the actions of Aboriginal peoples. After the demise of the Australian megafauna soon after humans arrived, the use of fire by humans in many instances continued to maintain biotic regimes established by large grazing herbivores such as *Diprotodon* species.[3] Conversely, the ease with which horses and other large introduced herbivores colonized parts of Australia may suggest that unoccupied grazing niches were available to them.

After escaping human control in the EM-R bioregion, brumbies became widespread and numerous during the late nineteenth century through to the 1970s. They were able to occupy habitats along the coast and inland from west of Esperance town to the east and north beyond Cape Arid. However, since the 1970s, as humans continued appropriating space and resources from nonhuman nature, their numbers have plummeted as vast tracts of land were cleared and fenced for industrial agriculture and other areas were alienated into parks and reserves. Brumbies are now reduced to two small populations—one of about thirty to forty in less than 200 square kilometers in the well-watered coastal Cape le Grand national park; the other inland in the arid eastern mallee-woodlands, where fewer than fifty or so may be left. Nevertheless, their continued presence and the purposeful trail pads they have developed over time reflect ownership and agency in their country and how in the past they successfully colonized the EM-R bioregion. In this chapter, I primarily refer to the colonization of the eastern mallee-woodlands rather than the Esperance sand plain populations because the history, culture, and ecology of brumbies in the mallee-woodlands is better documented and more visible than those of the horses that have long disappeared from the Esperance sandplain.

In this chapter, I argue that brumbies have culturally adapted to the mallee-woodlands and shaped the ecology in ways that maintain landscapes shaped by earlier herbivorous megafauna and Aboriginal inhabitants. In this way, the history of brumbies in colonial and postcolonial Australia suggests that in certain instances the introduction of domesticated animals from Eurasia into the so-called New World can be understood as a type of continuity rather than an abrupt change brought by the onset of European colonialism. Since introduction, brumbies became intimately engaged with landscape details and resources as they physiologically and culturally moved beyond their history of domestication and created what I describe as a "brumby cultural landscape." This is reflected through the knowledge of resources they use to find food and water, and through the creation of horse trails as they move purposefully throughout their home range. Yet there are signs that, like the decline of megafauna and the practice of Aboriginal "fire stick farming," the brumby landscape is in decline as a result of the increasing utilization of land for agriculture, human settlement, and protected areas that seek to "restore" ecosystems to a fabled and often idealized pre-European colonial condition.

Landscapes and Brumbies

The EM-R bioregion is located on the southeastern coast of Western Australia, and is part of the Nyungar, Wudjarri, and Ngadjunmaia traditional lands that form part of the Shire of Esperance.[4] At the coast, the bioregion features white beaches, turquoise blue ocean waters, and over 200 rocky islands forming the Recherche Archipelago. Inland there are landlocked islands, granitic rocky domes that rise above the flat to gently undulating plains as they did above the ancient shallow seas over 30 million years ago. These granite islands are important oases, providing water and growing lusher grasses and other vegetation within the surrounding dry mixed woodlands, bushlands, and grasslands. They have been a resource focus of native animals and Aborigines in the past, and later Anglo settlers and their animals. For brumbies, they form landscape nodes of food and water interconnected physically and culturally with their networks of trails.

The biogeographical ecozone inhabited by the brumby mobs (the Australian term for a group) that I will be discussing is part of the uncleared southeastern reserves and unallocated crown lands beyond Esperance agricultural land. They form the southern and eastern portion of the Great Western Woodlands and are dominated by large *Eucalypt* species interspersed with areas of *Mallee* species and mosaics of grassy herbaceous clearings. Comprising almost 16 million hectares (approximately 39 million acres), this is the largest relatively intact temperate woodland left on Earth.[5] Small bands of wild horses, descendants of the horses brought into the area by early European settlers, live in the woodlands and trek between areas of resource richness. Brumbies face threats such as dingoes (which can kill foals), human hunters, drought, and the occasional broken leg from slump holes in the limestone country.[6] They persist in these landscapes despite these challenges.

Ecosystems Reflect Animal Agency and Culture

Environmental history tends to emphasize change due to *human* influences because of the abundance of records produced by humans and the dangers associated with environmental determinism. There has been renewed interest in giving agency to nonhumans and more-than-human systems from a variety of fields. This research is informed by the recognition that the world's biophysical, biological, and nonhuman cultural interrelationships began well before the appearance of humans.[7]

A number of disciplines and researchers now accept that culture is widespread and not unique to humans. Research by animal behaviorists, such as Temple Grandin, provides evidence that animals using their superior senses such as sight, hearing, and smell perceive the world in images and details far beyond the range and comprehension of humans.[8] Particular forms of animal culture and language likely evolved as a result of the need to pass on specific survival details important to their particular population and species, such as observations about their habitats and ecosystems including predator avoidance, what foods to eat, and where water is found.[9] Animals, in this sense, create cultural relationships with particular home ranges based on continued interaction with and shaping of food resources, competition, and landscapes. In many instances, preexisting animal cultures, such as the creation of animal pathways, informs human settlement patterns.

The commencement of the present Ice Age epoch (Quaternary) 2.58 million years B.P. (before present) created environmental conditions that stimulated the development of gigantism in animals worldwide. In Australia and elsewhere, this suite of fauna has been termed "megafauna" (animals over 40 kilograms), though in comparison to other places a relatively small number of Australian species reached the supersizes (1,000 kilograms or more) common on other landmasses. Up until at least 50,000–41,000 years B.P., at least 340 species of land mammal (the majority marsupials) inhabited every suitable ecological niche in Sahul-Australia, as well as species of giant Varanid lizards and giant land birds known as *Genyornis* species.[10] Within the relatively short time span of several thousand years after the arrival of *Homo sapiens*, sixty-seven of these species became extinct.[11]

Before the arrival of people, the ecosystem evolution and resilience of Sahul-Australian landscapes was a function of the cultural and biophysical behaviors of the organisms that lived in them. Landscapes were engineered both purposefully and inadvertently by plants, animals, and microorganisms that maintained habitats and ecosystems in complex interacting and resilient webs of life, in ways that sustained individual species and maintained a collective ecology. Jones, Lawton, and Shachak have developed the following definition of how organisms, particularly herbivores among mammals, can be defined as ecosystem engineers: "Ecosystem engineers are organisms that directly or indirectly modulate the availability of resources to themselves and other species, by causing physical state changes in biotic or abiotic materials. In so doing they modify, maintain and create habitats."[12]

The concept of "keystone species" adds further to this paradigm as it proposes that there are some species that play a disproportionately important role in their ecosystems' structure and function.[13] Therefore, the evolution and resilience of pre-human Australian ecosystems was a function of the agencies of the organisms that lived in them. Organisms including plant species, the smallest microbes, and fungi, to the largest herbivores and various keystone predators, also have and have had major roles as ecosystem engineers.[14] Following from these paradigms it can be argued that the ecosystems of the EM-R bioregion are the result of interactions between the geological and living components of the landscape, which have over immense periods of time been undergoing adaptive cycles, uninterrupted by the geological upheavals that occurred in much of the northern hemisphere, and developed ecosystems with relatively long-lasting ecological resilience.[15] A large part of this adaptability and ecosystem resilience is reflected in the ability of living organisms—particularly large herbivores, including wild horses—to influence biophysical conditions in ecosystems and so improve their ongoing ability to exist and prosper.[16] In this way, brumbies may be important in maintaining the resilience and biodiversity of present EM-R bioregion ecosystems that are no longer managed by the Aboriginal social ecological systems that replaced the previous animal landscapes.

Bone deposits found in Western Australia near Balladonia, as well as caves on the Nullabor Plain and at Mammoth Cave near Margaret River, confirm the suite of wildlife along the south coast as similar to that found throughout Australia in similar environments. There were giant flightless Dromornithid birds (such as *Genyornis*) filling the giraffe niche; short-faced kangaroos that reached 3 meters in height may have filled a large primate niche; and the rhinoceros-like *Diprotodon*, and pig-like *Zygomaturus*, as well as large and small wombats. The wide diversity of kangaroo species were equivalents to the antelope and equine niche. Their predators included *Thylacoleo* the marsupial lion, thylacines (*Thylacinus cynocephalus*), and Tasmanian devils (*Sarcophilus sp.*). The largest predators were giant lizards up to 7 meters long, the *Varanus* (related to the present-day Komodo dragon).[17]

It is hypothesized by researchers such as, for example, Johnson and Prideaux et al., that large grazing herbivore species were vital in shaping ecosystem vegetation communities through their targeted herbivory that drove the evolution of plant morphology, such as spines and growth forms that protect the plant from grazing, as well as edible fruits to facilitate seed dispersal, along with physiological adaptations such as poison and unpleasant taste.[18] At the habitat level, mega grazers would have had a variety of impacts, including: reducing the density of vegetation and creating gaps and grassy mosaics that smaller herbivores could graze upon; increasing plant diversity; suppressing some species and favoring others; dispersing seed and facilitating niches for the animal coinhabitants of ecosystems; and reducing the frequency and intensity of lightning-ignited fires in the landscape by reducing accumulation of plant material and accelerating the rate of recycling and redistribution of ecosystem nutrients as they produced large amounts of dung and urine.[19] Large keystone predators such as the *Varanus* lizards and *Thylocoleo carnifax*, the marsupial lion, would also have exerted top-down control as a trophic cascade by regulating herbivore and smaller predator densities, which in turn affected plant community structure and biomass and then animals dependent on the configuration of these.[20] These keystone and ecoengineering species are proposed to have even exerted influence on geomorphology and climate as they created particular vegetation communities.[21] The biophysical processes described were integral in the creation of various feedback loops and influences that maintained ecosystems in a resilient state. At the level of individuals and populations, one of the most important influences upon ecosystem structure and function was through the agency of animal social systems and their cultures.

Many researchers define "culture" as the process of socially transmitting skills and knowledge to others.[22] Until recently, culture was considered to be an exclusively human trait and the concept that animals also created landscapes and interacted with them culturally was considered a step too far. For example, in his 1992 paper "Animal Culture," Galef examined the research at that time and concluded with a series of complicated reasonings that there is no evidence of animal culture that could not be explained by processes of instinctive behavior brought about as genetic transmission through natural selection.[23] Those who disagreed point out that transmission of information in this manner can only happen once during an individual's lifetime and its transmission by chance would mean that an enormous amount of time would be needed before it became a population trait. This time lag is inherently inefficient in contrast to the speed of cultural transmission that allows new behaviors to be learnt by many organisms in a matter of days and hours, rather than the many years of randomized natural selection it would take for behavior to spread among populations through genetic transmission.[24] Learning is a feature of all living organisms, which means that individual organisms can quickly achieve adaptation. This learnt adaptation can then be socially transmitted to offspring and others in the population through learning rather than genes, thus becoming culture.[25] The ability to pass on and learn new information may in itself become part of a population's evolutionary genetic makeup that favors the immediate adaptive flexibility of cultural behavior. Brumbies provide an example of this. They quickly learned how to access food and water resources in their landscape. Brumbies formed pathways to water and good grazing as a map-like landscape that future generations maintained through occupation. This knowledge is then passed on through generations based on the raising of foals by their mothers and groups.

During the Pleistocene, many of the Australian landscapes and vegetation communities were similar to those of the present. Large tracts of the EM-R bioregion would have been semiarid scrublands and woodlands as they are today, but featuring a far greater mammalian biodiversity than was present when Europeans first came to the region.[26] Fossil records confirm the presence of at least two species of the herbivorous wombat-related *Diprotodon* that can be envisioned as roaming through the EM-R bioregion grasslands in herds or family groups, and perhaps preyed upon by *Thylacinus sp.* and *Thylocoleo carnifex,* the marsupial lion. Many smaller marsupial predator and prey species were also present.[27] These herd-living *Diprotodon* species, and groups of giant macropods whose fossils have also been found on the south coast, may have lived in and maintained ecological niches analogous to those now occupied by brumbies.[28]

Paleontological evidence suggests that Pleistocene horse species, including *Equus ferus,* originated in North America and migrated to Eurasia via the Bering land bridge. They and many other megafauna became extinct in North America within a period coinciding with the arrival of humans approximately 13,000 to 14,000 years ago.[29] Throughout Eurasia they were reduced to the extant single genus by extinction events, with those wild equids left most diverse in savanna faunas in Africa and Eurasia, where today they fill the relatively restricted niche of large grazers of poorer fibrous plant material. Warmuth et al. provide evidence that the process of horse domestication was relatively recent and initiated about 450 generations B.P. Assuming a generation time of twelve years for wild horses, this corresponds to a start date for horse domestication around 6,000 years ago—a short time in which to evolve significant adaptations to a domesticated diet, which may explain why so many domestic horse illnesses are related to dietary factors.[30] It may also explain how brumbies in the EM-R bioregion and other places so readily returned to ancestral behaviors and food sources but may differ in some physiological traits compared with related African species that primarily rely on poor fibrous material. Feral horses seem to have expanded this typical niche because they include softer palatable browse and herbaceous plant species in their diets.

How Brumbies Became Colonizers

The Sahul/Australian animal landscapes were claimed by humans around 40,000 to 55,000 years B.P., leading to the demise of the megafauna and many other organisms that were codependent upon their ecosystem engineering and keystone species roles. Humans became the dominant cultural keystone species in terms of predation and also ecosystem engineering as they managed ecosystem plant communities, mainly with fire. Within these human socioecological systems, surviving species that could coexist also had vital roles in maintaining the resilience status of the new regimes. For example, the small soil-improving mammals included echidnas (*Tacchyglossus aculeatus*); burrowing bettongs (*Bettongia lesueur*); brush-tailed bettongs (*B. penicillata*), which have been measured as individually turning over up to 6 tons of soil in a year; potoroos (*Potorous sp.*); bilbies (*Macrotis lagotis*); bush rats (*Rattus fuscipes*); and southern brown bandicoot (*Isoodon obesulus*), which during their foraging and burrowing behavior turned over plant litter, dug shallow and deep holes, and spread seed and mycorrhizal fungi. This activity improved soil organic matter breakdown, provided a seed bed, and allowed rainfall absorption as well as perpetuating food resources for the species involved.[31] Kangaroos (*Macropus sp.*) and emus (*Dromaius novaehollandiae*) were the largest nonhuman animals left and because they were very important food for people, a significant amount of Aboriginal eco-farming with fire in the EM-R bioregion aimed to provide them with productive grassland habitats.[32]

The resulting mosaics of grassland were described positively by explorers such as John Septimus Roe as being "much good and available country, both arable and pastoral, has been seen in patches adapted to limited operations."[33] These grasslands were the initial attractant for the Anglo colonial pastoralist incursions that commenced in the 1860s and destroyed Aboriginal social ecological systems within twenty to thirty years. The settlers used the inherent biological traits of their domesticated animals, including horses, as an agency of colonization, a process described by Alfred Crosby as "ecological imperialism." Horses escaped to become wild, and as long as water was available, they were adapted both culturally and biologically to prosper in long-unoccupied ecological niches. Wild horses were the first large animals to live in the EM-R bioregion since the demise of the megafauna.[34]

Anglo settlers had distinctly different social, economic, and political systems to the Aboriginal inhabitants. They prioritized transforming lands they viewed as "dormant" into productive landscapes to produce goods for the monetary exchange economy. European settlers brought with them domesticated animals and plants that they were long familiar with as culturally appropriate food sources. In the EM-R bioregion, the transformations of Aboriginal social ecological systems that took place as a result of European colonization discarded the local animals and plants as food sources and were based first on pastoralism and later agriculture.[35] Therefore, Anglo occupation was interwoven with their culturally desirable domesticates. Like their masters' horses, cattle, sheep, and pigs, vegetables, cereals, and fruit plants had successfully extended their worldwide range through participation in previous European diasporas.[36]

Horses in Anglo colonial culture were rarely used for food. Horses functioned as the power source for long-distance transport of people and goods, and they provided the power for farm equipment. Horses pulled implements to plough the soil and seed the first cereals planted in the region—wheat grown to provide them with chaff and grain so they could continue to power colonization.[37] Horses had the ability to eat both native and naturalized plants like domestic ruminants.[38] However, because they are not ruminants, they were largely immune to the mineral deficiency diseases that plagued introduced ruminant herbivores along the southern Western Australian coastal lands. Ruminant herbivores needed to be taken inland to the better soils for a certain period each year to prevent the coastal disease caused particularly by lack of the trace elements cobalt and copper, as well as the macro element phosphorous. Horses were used to drove cattle and sheep between the coast and inland for this practice of transhumance.

As strongly social animals capable of forming deep attachments to each other, horses are also capable of forming strong attachments to people who treat them with respect and firmness and so were often viewed with deep affection by their human masters.[39] This appeared to be an important part of the emotional structure and interactions of colonization using horses. The first horses recorded in the EM-R bioregion were those of the explorers. Their journals confirm that often there were strong emotional attachments. Starting from Perth in 1848, John Septimus Roe explored the mallee and woodlands north and east of Esperance, then returned along the coast. The affection for his horses is confirmed by the effort expended to keep his favorite horse "Ney" alive. Much time was spent coaxing the exhausted animal to a large granite outcrop with plentiful grass and water where he could recuperate: "As both grass and water were abundant at this limited spot, I determined on leaving him here to have a chance of recovering from his exhaustion, and of being called for again on our return homewards by a more southerly route."[40] Later, when on the coast and about 60 kilometers south of where Ney had been left, two days were put aside to go and fetch the horse. The granite outcrop where he had successfully recovered was officially named Mt. Ney by Roe in his honor.

This human–horse affection theme is continued among other colonial settlers. In 1874, the educated and rather genteel Brooks family from Victoria, Mrs. Emily Brooks, her son John Paul, and daughter Sara Theresa walked around 800 kilometers from Albany to Thomas River in the EM-R bioregion in response to the government's promise of the availability of large tracts of land to lease. However, by the time they arrived most of the land was taken. So, leaving his mother and sister at Thomas River Station, John Paul and a party of men and horses set out for Eucla to look for land. During the expedition, his horse Jessie was left behind after they all nearly died of thirst. He spent many hard days searching before finding her: "Home again, hurrah! . . . The magic effect on poor Jessie, and myself at the sight of the horses, the cart, and the tent, could be caused by nothing less than home. . . . Nobby [another horse] and Jessie are kissing one another at a great rate. . . . J.D. says 'Well, my goodness! Mr Brooks, there's no mistake you deserve her, there's not many men in this country would have walked 120 miles for her, but she's all right now, and she'll live to pay you, I would have staked my life you would not have found her alive.'"[41] This family eventually settled at Balbinya station, inland from Israelite Bay.

The Dempster brothers were the first Anglo settlers to be given leaseholds in the Esperance bioregion in 1863, so commencing the destruction of landscape management systems of the local Aboriginal peoples. Horse breeding was important to the family farm business at Toodyay, taken up in 1843 near Perth, the capital of Western Australia. Here young horses were trained and sold to India for the remount market.[42] The brood mares ran semiwild in the bush after mating but could be located if needed because the home range of each horse was known. Mustering to catch weanlings took place at the end of spring, when horses congregated at the few remaining waterholes.[43]

In 1864, few of the Dempster horses were sent to India. Instead, some pulled wagons to transport stores and equipment to Esperance Bay and others were ridden to drove the cattle and sheep flocks to the new properties. With an unfenced pastoral empire of over 600,000 hectares, horses were vital as mounts for the shepherds who took stock from place to place following the feed and water. They were also necessary to effect the local transhumance system used to avoid coastal sickness in sheep and cattle. Though the Dempsters would not have known the underlying cause (as previously described), earlier observations elsewhere in the state had found the solution was to move stock to mineral-rich inland pastures. Sheep and cattle were walked from Esperance 200 kilometers inland to Fraser Range and back every year. Historical records are ambiguous about horse escapes or releases by the Dempsters. With the large numbers owned in such a huge area and their probable continuation of the practice of letting mares run in the bush, it is very likely that some may have escaped beyond human influence to become brumbies.[44]

In 1901–2, Heinrich Dimer (a German sailor who jumped ship in 1884) and his part-Aboriginal wife Topsy, were granted over 25,000 hectares of pastoral leases north of Point Culver, which became Nanambinia Station. Their main sources of income were from wool and cattle, with station management based upon Aboriginal shepherds moving flocks of sheep and some cattle between the scattered granite outcrops and following the rain. Without permanent streams and no rivers in the mallee-woodlands, the Aboriginal-made gnammas and small catchments on these granite outcrops were the only water source.[45] Water was the ongoing concern because these water storage systems held too little water to provide for large numbers of domestic animals. Using camel and horse-pulled scoops, over time the Dimers built a number of large dams and improved catchments around the numerous granite outcrops to increase the amount and permanency of water on their properties. This provided the water base for running large numbers of sheep, cattle, donkeys, horses, camels, and wildlife.[46] Other settlers also used this strategy so that an attractive feature today of most granite outcrops in this drier region is the deepened and enlarged run-off dams around their bases that create almost permanent water storage.

This provision of water to both stock and wildlife (including animals that later became wild, such as horses) was a fundamental ecological impact of pastoralism in this part of the EM-R bioregion. Although this allowed the carrying capacity to be increased, it also made overstocking more likely by wild as well as domestic herbivores. Even so, erratic rainfall and drought, especially from the 1930s to 1940s onward, meant that water continued to be a limiting factor. As Karl Dimer notes in his history of life on the station, "the scarcity of water, in fact, tended to rule our lives." This intermittent water scarcity is likely to have prevented the level of overstocking that probably occurred on the water-abundant coastal areas leased by others.[47] In present times, lack of water during droughts is an important restrictor of brumby population growth in the eastern mallee-woodlands as some, especially mares and foals, die of thirst.

Like many settlers, the Dimer family had a strong relationship with their horses, both as an economic necessity for income and for managing the large areas of land they leased, but also at an emotional level that was built on affection and respect for the individual horses as personalities in their own right. Horses were bred for stockwork, with some sent to a business partner who broke them to saddle for sale as saddle horses. All were named and known as individuals and used in rotation to rest them physically and emotionally, sometimes as saddle horses, or pulling the sulky, cartage wagons, and ploughs. When given time off they were freed until needed again or when mustered to castrate young males and handle and train yearlings. As Karl describes, individuals could always be found on the huge property because the home ranges to which each consistently returned were well known: "I went off on a mare called Darling . . . to look for some saddle horses. Each mob of horses had their own territory in which they liked to run. . . . I was looking for Ringer who was a very good saddle horse, and I knew he always ran at a place called Gudamoona, which was his birth-place."[48]

A horse would travel great distances to return home. Karl let a mare go 100 kilometers south of Nanambinia at Mt. Ragged: "I let Rachel go, knowing she would go back to her own run, or district, and that next time we wanted her, we would find her there."[49] So strong is this homing desire in horses that they would attempt even greater distances to return home. During the late 1920s, agricultural farming started up at Salmon Gums and Eastern State Shire horses were imported to chain and plough the virgin bushland for agriculture. A number escaped and tried to return to their homes—over 1,000 kilometers from Salmon Gums; with the arid, almost waterless Nullarbor Plain to cross, many must have died in the attempt. The Dimers rescued a number of them and rehomed them on their station. Their genetic influence can be seen in some mobs of present brumbies with their heavier build, hairy feet, and white facial blazes.[50]

Karl described how the horses loved their freedom and had many pads (trails) going to so many places, you had to be right up with them and know where the pads went to catch them as they could dash down a pad and get away: "To prevent the horses getting away . . . you had to turn off a mile or so before the pads and be waiting there when the horses came along it. You could just about see the surprise on the horses' faces when they confronted a rider on what they thought was their getaway chance."[51] Some were not caught. Windjammer may have been one of the early brumby ancestors because she was never caught: "Syd tried his luck with Windjammer, who got the name because she was so fast, but he had no luck in getting her in either."[52]

Gradually, pastoralism at Nanambinia petered out and during the 1940s the partnership was split, the Aboriginal workers translocated or died and "[the] horses back westward were left to run wild" and so joined the ancestors of today's brumbies. Since they already knew the country and most had been born there and grown up in it, the transition from domestication to wild would have been seamless.[53]

Freed of human control and successfully filling a vacant ecological niche unoccupied since the demise of the megafauna, brumbies became widespread and numerous, roaming along the coast and inland from west of Esperance town to east and north beyond Cape Arid. Until the 1970s, they were accepted in the landscape as they came into the town to graze on road verges, lawns, and in the town parks. Their long-term demise started as vast tracts of sandplain and mallee-woodlands land were alienated for agricultural development after the Second World War. As did native wildlife, their numbers plummeted as large-scale clearing and fencing commenced from the 1950s to the 1980s. Until clearing bans were introduced in the latter part of the 1980s, about 1,000,000 hectares of land were eventually developed for agriculture and brumbies and native wildlife lost their habitats and either died of starvation and predation or were destroyed by edicts from government National Parks bureaucracies, which showed an unrelenting intolerance toward anything termed "feral." There are now only two small populations left in the EM-R bioregion—one in Cape Le Grand National Park and the other in the mallee-woodlands east of the farmland.

Brumby Mobs and Landscape Resources in the Eastern Mallee-Woodlands

Ungulates, including wild horses, rarely range randomly. They have places they prefer and others which they avoid. Though no long-term ecological studies have been attempted with the EM-R bioregion group of horses, observations over a number of years have indicated that parameters of habitation appear similar to those found in studies elsewhere by researchers such as Patrick Duncan.[54] These indicate that composition and structure of grazed plants and their nutritional variation throughout the year is important in determining where horses will move to; another factor that Duncan has shown to be important is the presence of biting insects, particularly blood-feeding March flies (family Tabanidae—called horse flies in other countries). They are present in large numbers during late summer to autumn and found in the shady forested areas; and if behavior correlates with Duncan's findings, horses would avoid feeding in such areas at times when these flies are most abundant.[55] In this part of the EM-R bioregion the occurrence of successive years of low rainfall and drought and the presence and permanency of water and remaining feed is a major factor in brumby spatial organization. Studies elsewhere in Australia have found the size of the home range directly related to yearly food and water availability.

In central Queensland, a study of brumby movements by Hampson et al. used GPS trackers to record the distances and locations traveled to by two populations of brumbies in different resource areas. They found that horses traveled as far as 65 kilometers from watering places and would go up to four days without water in their search for food in the less hospitable landscapes. This ability provides evidence that without human intervention for the past 150 years, brumbies may have started to evolve physiological adaptations similar to those of their never-domesticated wild cousins, Przewalski horses (*Equus ferus przewalskii*), to improve water use efficiency and withstand dehydration more easily than their domestic ancestors. Brumbies, when dehydrated, are recorded as being able to drink large quantities of water in a few minutes and rehydrate rapidly.[56] In the EM-R bioregion, brumbies also typically travel far greater distances daily than their domesticated counterparts and eat a far wider selection of high-fiber foods as well as nutrient-dense foods. Domestication has resulted in confinement and altered feeding patterns, which in many ways parallel the modern human lifestyle that is thought to be responsible for endemic levels of obesity, cardiovascular disease, arthritis, and diabetes. A common affliction of domesticated horses is Equine Metabolic Syndrome linked to laminitis. This disease is analogous to human diabetes as it is caused by eating high sugar (fructan) energy feeds such as found in lush pasture and grain with inadequate roughage, which can lead to obesity and insulin resistance.[57]

In the mallee-woodlands, native and naturalized grasses, succulent herbs such as pigface (*Carpobrotus spp.*), and bluebush and saltbush shrubs are scattered patchily throughout the landscape. The mix of soil types and scattered granite outcrops ensures numerous vegetation communities, including tall eucalypt woodlands with an understory of grasses and edible bluebush (*Maireana spp.*) and saltbush species (*Atriplex spp.*). There are also woodlands of native cyprus (*Callitris spp.*), many species of mallet, mallee and banksia shrublands, and Kwongan heathlands. North of Israelite Bay there are numerous treeless grassy glades and alkaline karst plains (such as found at Kangawarrie) on stiff clay soils underlain by limestone and featuring underground caverns and slumps. The treeless clearings and granite outcrops grow many edible native plants such as perennial kangaroo grass (*Themeda triandra*), wallaby grass (*Austrodanthonia caespitosa*), and spear grasses (*Stipa eromophila; Austrostipa nitida*); there are bi-annuals such as windmill grass (*Chloris truncata*), herbs including pigface, and shrubs such as bluebush and saltbush.[58] Numerous feral plants have colonized including blowfly grass (genus *Briza* from the Mediterranean), medic legumes (*Medicago*, also of Mediterranean origin), and South African capeweed (*Arctotheca calendula*), which all contribute to edible species. There are also some very poisonous native plants that brumbies have somehow learnt to avoid, such as wild tobacco (possibly a *Solanum sp.*) and *Gastrolobium sp.* (which contains the toxin sodium fluroacetate used in the manufacture of 1080 poison) that grow around and on granite outcrops. Native mammals can tolerate this poison, but it is fatal to introduced herbivores and predators. Hypothetically, avoidance is likely to have somehow been a socially learned behavior that has become an intergenerational tradition, perhaps through distant past observation of others dying after eating this plant. The invariable and rapid death from these plants does not allow for an individual to learn avoidance from mistakes. Brumbies may have also evolved a level of immunity to this toxin, although this also is an untested hypothesis.

The best grazing areas for wildlife, including brumbies, are found around the granite outcrops. Rain runoff promotes fertile, moist mini-habitats, encouraging introduced and native edible grasses and other plants to thrive. With no creeks or rivers, water resources are also focused on the granite outcrops, which catch water in the ancient Aboriginal gnammas and deep rock depressions formed through natural weathering. After good rains, large transient puddles occur along the few dirt roads. Largely permanent water is found at Pinehill, which has a large runoff dam deepened and extended by settlers. Breeborinia Rock, 25 kilometers to its northwest, also has a permanent deepened runoff dam. A number of other outcrops to the north, northwest, east, and northeast have semipermanent to mostly permanent waters, some in gnammas and others where Anglo settlers constructed clay dams and rock walls on granite to funnel water into small dams scooped out around the edges. All are connected by brumby pads, with the most used being those that lead to the most permanent water sites.

Social Systems and Land Ownership by Horses

With the abundance of edible plants and improved waters, horses were biologically preadapted to the mallee-woodlands of the EM-R bioregion. Their social and cultural behaviors that allow transmission of knowledge and the reduction of conflict and inbreeding are also adaptive. These factors ensure that knowledge is passed on to offspring, who take it with them when forming new mobs and ensure unnecessary competition for space and mates is minimized.

Linklater et al. have determined how horses interact socially and spatially. They have found that both sexes tend to disperse from their natal bands so that matrilineal groups are not formed. Fillies will join a mare band, while colts form transient bachelor groups after dispersal. Horses form long-term stable bands of mares, with one to four usually permanent stallions (one stallion is sexually dominant, and the others have arrested sexuality). The bands are as small as one mare and one stallion, or up to twenty mares with associated stallions. The stallion(s) associated with the mare bands can also be long term. This type of society is termed a female defense polygyny. Home ranges overlap with bachelor groups as well as those of other bands, and there is little evidence of territorial defense. There is strong loyalty and attachment to core home range and strong friendships can form between individuals.[59] In the mallee-woodlands most bands are small, and brumbies do not overpopulate because there are significant constraints on their numbers. In particular, finite water availability means death from starvation and thirst during the uncommon but inevitable serious droughts, where even "permanent" waters dry up and feed disappears. Dingoes may take foals and horses are injured by environmental hazards such as karst plain slump holes, where broken legs can occur. Recently, more are being killed by human hunters who chase them with quad bikes and shoot them for fun. Their indiscriminate shooting of females and foals could eventually lead to local extinction.[60]

For past and present animals and humans living in this region, being familiar with and knowing the exact location of life-sustaining landscape features is a necessity for survival. Traveling between the scattered granite domes, which are not only a resource base but also navigational markers, brumbies visibly occupy and claim areas of the mallee-woodlands. Their crisscrossing trails and stallion dung piles left at strategic points, illustrate how deep past large animal landscape interactions may have looked. Tracks and trails are a physical manifestation of many animal cultures, a way to promote memory through association that helps individuals to find important places. Locations of resources are learnt from mothers and others as targeted rather than random movements.[61] The web of brumby footpads trodden for generations connects the important landscape nodes of food and water. Perhaps these trails originally followed Aborigine and/or kangaroo tracks, but many were certainly horse tracks, developed as areas of importance to them were learnt and linked in space and time. Research elsewhere has also indicated that horse trails were used frequently by other wildlife, such as emus and kangaroos, for travel routes.[62]

Breeborinia rockhole and Pinehill dam are about 20 kilometers apart and the most direct route between them is across a huge gypsum lake system featuring treacherous gypsum dunes, scattered bogs, and quicksand. There are scattered sidetracks to other watering places and feeding areas and the linking tracks at first appear to meander aimlessly. Instead, they actually find the safest ways to cross the lake. My colleagues and I found this out on an exploratory expedition to work out how far brumbies travel between two waters by riding our horses from a camp near Breeborinia to Pinehill and back in one day (over 40 kilometers).

The horse trails meander across only the eastern portion of the lake system. Failing to question why, we decided to cross at the western end, which was a shorter distance from our camp to Pinehill. After sinking into quicksand, it became obvious that there were so many hazards to avoid such as quicksand, bogs, and deep dry gypsum dunes that our trip would take far longer than estimated from using maps. It took many dangerous and exhausting hours to finally get through. On the return journey we determined to cross at the eastern end following brumby pads. Riding along them it became obvious that they deliberately avoided the hazards and because the trails had generations of use, they had formed into compacted narrow tracks much firmer than the lake beds they passed over. Our horses did not instinctively understand this, and it took a few unpleasant boggy episodes off the trails before we and they learnt that it was best to only walk on the trails. In spite of being longer, it took far less time to return than our outward journey as we acknowledged that brumbies are cultural beings with a deep knowledge of their homelands.

Brumbies: "Feral" Landscape Destroyers or New Ecosystem and Landscape Engineers?

The presence of brumbies in the Australian landscape, including the EM-R bioregion, is controversial to a significant proportion of Australians who perceive them as environmentally damaging feral animals. The use of the word "feral" has become politicized as an implicit justification to aim for elimination of nonnative wild animals throughout Australia, including wild horses. There is much discussion about management of wild horses based on perceptions of their environmental impacts, which in this region are based on research conducted elsewhere as there have been no empirical studies that accurately represent their agency in local ecological systems. Though seen as feral pests by most land managers, there is a growing minority in the local community with a more "kincentric" worldview, where the agency of horses (and other wildlife) is recognized as having become a legitimate part of local cultural, historical, and ecological systems.[63]

Among the brumbies themselves during the last 160 years of occupation of the EM-R bioregion, their behavioral relationships within their homelands have moved them beyond the biophysical into the realms of cultural landscapes. This is supported throughout the discussion by Albrecht et al. about the rights for existence of "feral" buffalo in northern Australia, where the concept of "ecosystem being" is introduced. This view is based on the idea that a species and its individual members have over time developed an identity integral to their relationships with elements of the ecosystems they occupy and that they are now a well-established component within a complex adaptive system, such that their removal may have a cascade of unforeseen and negative consequences.[64] Integral to this concept is recognition that the feeding behaviors of large mammalian herbivores can be important in determining the composition of plant communities, with their presence leading to increased plant biodiversity that has a complementary impact on animal biodiversity. They also recycle nutrients and reduce the flammability of landscapes by removing excessive vegetation dry matter build up.[65] From an Australian context with the extinction of the original large herbivore keystone species, fire was used by humans in Australia and the EM-R bioregion as a key ecological process for maintaining biodiversity and preventing wildfires.

As a consequence of postcolonial dislocation of Aboriginal people from land management in the EM-R bioregion (and throughout Australia), fire is no longer an integral part of a full-time cultural landscape management system. The current bureaucratic systems in place are limited by political and economic factors, lifestyle and workplace regulations, and different philosophies—none of which is proving effective in managing the outbreak of wildfires, which have such destructive impacts on people, property, and ecosystems. Catastrophic fire risk periods are also becoming more frequent, as human-induced climate change is resulting in reduced rainfall and increasing temperatures. This suggests that a complete cultural paradigm shift is required to economically and ecologically reduce the impacts of this devastating landscape shaper.[66] Ecological substitution with large herbivores such as horses, camels, and cattle can modify landscape flammability as they create mosaics through patchy grazing and can provide firebreaks that prevent small fires from becoming large and destructive.[67] They could be used in nature conservation to facilitate self-managing functional, biodiverse ecosystems (rewilding). The concept of rewilding landscapes with large herbivores to restore ecological function is an ongoing discourse that is being progressively implemented throughout the world, with the highly adaptable

E. ferus caballus an appropriate candidate.[68] For such a paradigm shift to occur in Australia, it would first be necessary to overthrow the cultural myth that the feet of all hoofed mammals have destructive effects upon Australian soils. Rangeland ecologist P. B. Mitchell asserts that the "common sense" myth of hard hooves as the primary cause of soil compaction and therefore land degradation in semiarid Australia is supported by little empirical evidence, but as an uncritical generalization, it has become rife in scientific and popular literature. Accordingly, not enough attention is given to potentially more important factors in land degradation such as the impact of excessive stocking rates, preferential grazing, and animal behavior.[69]

There is abundant evidence that where there is an overpopulation of brumbies they can damage or negatively change particular Australian ecosystems. However, there is also evidence that moderate levels of brumby grazing that reduce plant biomass and thus flammability, as well as create and maintain a patchy landscape, could facilitate positive effects on ecosystem biodiversity as predicted by the ecosystem disturbance hypothesis.[70] This hypothesis theorizes that moderate levels of ecosystem disturbance can enhance and maintain biodiversity in ecosystems.[71] I would argue that the effect of starting out with a research imperative that tends to have a preconceived view of ecological roles as harmful by default can be seen in much of the primary research completed to date in Australia, which has tended to search for particular situations where the focus is on negative impacts in small-scale localized situations, from which sweeping analogies are then drawn across large-scale landscapes. Inherent qualities of *E. ferus caballus* as a species never native to Australia are being assumed as the primary factor causing these impacts, rather than examining the link between overpopulation impacts where any species (including humans, sheep, and kangaroos) can overexploit and biophysically damage ecosystems. This is especially relevant in Australian ecosystems because the only native predator large enough to prey on horses and perhaps limit their population without human intervention is the dingo (*Canis dingo*). Since colonial times, this medium-sized predator has been the focus of high-level persecution and population suppression as it is viewed as an unwanted destructive predator within agricultural social ecological systems that have largely been based on sheep. Therefore, other than a few anecdotal instances, there has not to my knowledge been any serious consideration or research directed toward the role of dingoes as a keystone predator in controlling brumby populations.

Nimmo et al. concur with my general argument as they concluded from a review of Australian research that there are a number of serious gaps in understanding "the ecological effects of feral horses on native environments, particularly with regard to Australian ecosystems."[72] Until these understandings are reached through peer-reviewed research that examines brumby populations and their ecosystem interactions across a range of landscape scales and without the "harmful by default" view, then, as Wallach contends, the new Australian megafauna—including brumbies in the EM-R bioregion—"are, of course, changing their new habitats, possibly by bringing back [a] lost Pleistocene functionality."[73]

Conclusion

The history of brumbies in the EM-R bioregion offers an important alternative example to studies that see the introduction of Eurasian fauna as a type of "ecological imperialism" with devastating consequences. So too does the history of brumbies challenge contemporary purist notions of ecological restoration, which sees introduced species as problems to be managed or eradicated. Brumbies in the EM-R are examples of the hybridity of new ecosystems that have been created as a result of globalization.

Overall, during the last 160 years, brumbies as colonists have adapted well to the eastern mallee-woodlands of the EM-R bioregion. They have learnt to accurately locate water and feed and how to avoid eating poison plants. Intergenerational trail pads have been developed that are safe and efficient and link important food and water nodes throughout the landscape. As big animals, like many of the extinct megafauna and much larger than the native animals that had survived with Aboriginal social ecological systems, brumbies may indicate how past animal cultures organized landscapes. They may be considered as positive rather than negative agents of ecological change, capable of bringing back some of the past ecosystem functions of extinct megafauna that enabled resilience and sustainability through biodiversity, nutrient cycling, and fire mitigation. For instance, though untested by research, anecdotes suggest reduced wildfire frequency and intensity in the EM-R landscapes where they live, supporting evidence that prehuman animal landscapes in Australia may have been significantly less fire prone.[74] Brumbies also intimately know their landscape and the resources it offers as they move around in their homelands, behaving as do present megafauna in Africa and, by implication, how extinct Australian megafauna would have proactively occupied habitats in the EM-R bioregion. As yet there has been no research on the possibility that brumbies could have positive influences on the landscapes they live in as large herbivores. There is a plethora of negative discourse on their assumed deleterious impacts as the illusion of "natural and unchanging wilderness" in Australia is advocated. Yet this illusion is based on pre-European landscapes that were so intimately influenced and managed by Aboriginal social ecological systems that they were not a wilderness and were not the same landscapes that existed in prehuman Australia.

The following quote by Pat Fischer, Gooniyandi Traditional Owner, reflects a different way of conceptualizing the world and its ecological networks as a function of the creatures who inhabit it along with humans. From them we can discard human arrogance and learn: "Brumbies do not belong to us they belong to the land. They are teachers of natural ways. They tell us what the seasons will be like, when the dry times are coming. They find water in the dry river beds with their hooves, water that brings life to the cattle, native animals and birds. They are now part of our Country, they reflect its health. They deserve the same care and respect due to all in Creation. Allow them the vast open spaces, allow them the land to run free, honour them for what they are."[75]

Notes

1. Robyn MacDougall, *The History of the Guy Fawkes River Australian Brumbies and the Brumbies of the Northern Tablelands* (New South Wales: Self-published, 2001), 9; Frederick Ludowyk, *Wild Horses Running Wild: Chasing Our Brumby*, https://www.ausemade.com.au/fauna-flora/fauna/mammalia/perissodactyla/equidae/equus/brumby/brumby.htm.

2. Alfred W. Crosby, "Ecological Imperialism: The Overseas Migration of Western Europeans as a Biological Phenomenon," in *The Ends of the Earth: Perspectives on Modern Environmental History*, edited by Donald Worster (New York: Cambridge University Press, 1989), 103–17; Eugene N. Anderson, *Everyone Eats* (New York: New York University Press, 2005), 89.

3. Tim Flannery, *The Future Eaters: An Ecological History of Australasian Lands and People* (Chatswood, NSW: Reed Books, 1994); Chris N. Johnson, *Australia's Mammal Extinctions: A 50,000-Year History* (Cambridge: Cambridge University Press, 2006), 212–14; Matt McGlone, "Paleontology: The Hunters Did It," *Science* 335 (2012): 1452–53.

4. See terminology adapted from Department of Sustainability, Environment, Water, Population and Communities, *Interim Biogeographic Regionalisation for Australia, Version 7* (Canberra, ACT: Australian Government, 2012). Most groups were likely to have been multilingual, so interpretations by early anthropologists may have been inaccurate and circumstance dependent. The area is part of an area under claim by the Esperance Nyungars and the Ngadju people.

5. Alexander Watson, Simon Judd, James Watson, Anya Lam, and David Mackenzie, *The Extraordinary Nature of the Great Western Woodlands* (Perth, WA: The Wilderness Society of WA Inc./Scott, 2008). The Great Western Woodlands (GWW) is a 16 million hectare swath of grasslands, mallee, woodlands, and heathlands interspersed with salt

lakes, which represents the largest intact remaining Mediterranean woodland habitat in the world; http://www.gondwanalink.org/whatshapwhere/gww.aspx (11 February 2017).

6. Trent Ridgway, personal communication, 2013; the Ridgway family has a lease over a huge lake system for gypsum mining and often see wild horses—sometimes in the hazardous, hole-ridden, limestone flats, sometimes with broken legs.

7. Stephen Dovers, "Sustainability and 'Pragmatic' Environmental History: A Note from Australia," *Environmental History* 18, no. 3 (Autumn 1994): 21–36. In this article, Dovers explains that "In seeking a sustainable relationship between human and natural systems we must first construct histories, establish baselines and identify long term trends." A. R. Main, "Ghosts of the Past: Where Does Environmental History Begin?" *Environment and History* 2 (1996): 97–114. Main discusses the premise that modern landscapes in Western Australia are the result of deep past animal and plant adaptations, as well as Aboriginal land management practices; for instance, research indicates that small fungi-eating, digging marsupials such as bettongs are keystone species in soil structure and function. Greg Martin, "The Role of Small Native Mammals in Soil Building and Water Balance," *Stipa Native Grasses Newsletter* 16 (Autumn 2001): 4–7.

8. Kalevi Kull, "Adaptive Evolution without Natural Selection," *Biological Journal of the Linnaean Society* 112, no. 2 (2014): 287–88; Temple Grandin and Catherine Johnson, *Animals in Translation: Using the Mysteries of Autism to Decode Animal Behavior* (New York: Harcourt, 2006). Grandin is autistic and acknowledged as a leader in understanding how animals think and probably see the world. Her proposition is that autism allows her to see the real world in tremendous detail, visualized in pictures; and this is likely how animals, particularly mammals and birds, interact with their environment—totally necessary for finding food, mates, and avoiding hazards. "Normal" humans, in contrast, generally see an "idea" of the world as constructed by their forebrain through generalized experience.

9. Grandin and Johnson, *Animals in Translation*, 136–37, 292–93. Most mammals and birds must learn what to eat and what not to eat from their parent(s) because it is not an instinctive process.

10. R. Morrison and M. Morrison, *The Voyage of the Great Southern Ark: The Four Billion Year Journey of the Australian Continent* (Sydney, NSW: URE Smith Press, 1988), 267–69.

11. Johnson, *Australia's Mammal Extinctions*, 16–35; Susan Rule, Barry W. Brook, Simon G. Haberle, C. S. M. Turney, Arnold Peter Kershaw, and Christopher N. Johnson, "The Aftermath of Megafaunal Extinction: Ecosystem Transformation in Pleistocene Australia," *Science* 335 (2012): 1483–86.

12. Clive G. Jones, John H. Lawton, and Moshe Shachak, "Organisms as Ecosystem Engineers," *OIKOS* 69 (1994): 373–86. Following on from these assertions, human beings are definitely ecosystem engineers on a grand scale from the past to the present but the roles of other organisms are only recently being teased out.

13. L. Scott Mills, Michael E. Soulé, and Daniel F. Doak, "The Keystone-Species Concept in Ecology and Conservation," *BioScience* (1993): 219–24; Kevin N. Laland, John Odling-Smee, and Marcus W. Feldman, "Niche Construction, Biological Evolution, and Cultural Change," *Behavioral and Brain Sciences* 23, no. 1 (2000): 131–46;

Mike Letnic, Feya Koch, Chirs Gordon, Matthew S. Crowther, and Christopher R. Dickman, "Keystone Effects of an Alien Top-Predator Stem Extinctions of Native Mammals," *Proceedings of the Royal Society B: Biological Sciences* 276, no. 1671 (2009): 3249–56, which examines the role of dingoes in ecosystem structure and biodiversity.

14. Eric A. Smith and Mark Wishnie, "Conservation and Subsistence in Small Scale Societies," *Annual Review of Anthropology*, no. 29 (2000): 496–97. The ecosystem engineering concept is discussed in depth in the following two papers—Jones et al., "Organisms as Ecosystem Engineers," 373; Justin P. Wright and Clive G. Jones, "The Concept of Organisms as Ecosystem Engineers Ten Years On: Progress, Limitations, and Challenges," *BioScience* 56, no. 3 (2006): 203–20. Examples are provided by the following papers: Christopher N. Johnson, "Ecological Consequences of Late Quaternary Extinctions of Megafauna," *Proceedings of the Royal Society B: Biological Sciences* 276 (2009): 2509–19; P. M. Huang, "Foreseeable Impacts of Soil ML-Organic Component-Microorganism Interactions on Society: Ecosystem Health," in *Ecological Significance of the Interactions among Clay Minerals, Organic Matter and Soil Biota*, edited by A. Violante, J.-M. Bollag, L. Gianfreda, and P. M. Huang, (Amsterdam: Elsevier Science, 2002).

15. Hans Lambers (personal communication) has described how Great Western Woodland ecosystems are among the longest lasting and most "stable" on earth due to such long periods of geological stability in Western Australia. This is supported by the many plant species that directly modify the soil through pedogenesis, recently discovered by the following authors: William H. Verboom and John S. Pate, "Exploring the Biological Dimensions to Pedogenesis with Emphasis on the Ecosystems, Soils and Landscapes of Southwestern Australia," *Geoderma* 211–12 (2013): 154–83; Carl Folke, Steven R. Carpenter, Brian Walker, Marten Scheffer, Terry Chapin, and Johan Rockström, "Resilience Thinking: Integrating Resilience, Adaptability and Transformability," *Ecology and Society* 15, no. 4 (2010): 20. Resilience thinking has been developed as a way of understanding social ecological systems; the premise is that three aspects are central: resilience, which allows adaptability and transformability that interrelate across multiple scales, is just as true for all ecosystems.

16. Theoretical considerations of how organisms are believed to create habitat are found in Wright and Jones, "Concept of Organisms," 203–9. For examples of ecosystem engineering by plants, which the authors refer to as the *Phytotarium Concept*, see William Verboom and John Pate, "Exploring the Biological Dimensions to Pedogenesis with Emphasis on the Ecosystems, Soils and Landscapes of Southwestern Australia," *Geoderma* no. 211–12 (2013): 154–83.

17. P. F. Murray and P. Vickers-Rich, *Magnificent Mihirungs—The Colossal Flightless Birds of the Australian Dreamtime* (Bloomington: Indiana University Press, 2004); Patricia Vickers-Rich and Thomas Hewitt-Rich, *Wildlife of Gondwana* (Chatswood, NSW: Reed, 1993).

18. See the following for descriptions of megafauna species found in Western Australia, plus plant communities and adaptations: Gavin J. Prideaux, Grant A. Gullya, Aidan M. C. Couzens, Lynda K. Ayliffec, Nathan R. Jankowski, Zenobia Jacobs, Richard G. Roberts, John C. Hellstrome, Michael C. Gaganc, and Lyndsay M. Hatcherf, "Timing and Dynamics of Late Pleistocene Mammal Extinctions in Southwestern

Australia," *PNAS* 107, no. 51 (2010): 22157–62; Johnson, "Ecological Consequences"; Vickers-Rich and Hewitt-Rich, *Wildlife of Gondwana*.

19. David M. J. S. Bowman, "Australian Landscape Burning: A Continental and Evolutionary Perspective," in *Fire in Ecosystems of South-West Western Australia: Impacts and Management*, edited by Ian Abbott and Nial Burrows (Leiden: Backhuys, 2003): 113–14; Anthony R. E. Sinclair, "The Role of Mammals as Ecosystem Landscapers," *ALCES* 39 (2003): 161–76.

20. Letnic et al., "Keystone Effects," 3249–56; for cascading trophic effects due to removal of an apex keystone predator, see Tracy A. Hollings, *Ecological Effects of Disease Induced Apex Predator Decline: The Tasmanian Devil and Devil Facial Tumour* (PhD diss., University of Tasmania, 2013), 2.

21. Ingeburg Burchard, "Anthropogenic Impact on the Climate since Man Began to Hunt," *Palaeogeography, Palaeoclimatology, Palaeoecology* 139 (1998): 1–14; Rule et al., "The Aftermath of Megafaunal Extinction," 1483–86; George Monbiot, *Feral: Rewilding the Land, Sea, and Human Life* (Chicago: University of Chicago Press, 2013), ch. 7.

22. Carel P. van Schaik, "Social Learning and Culture in Animals," in *Animal Behaviour, Evolution and Mechanisms*, ed. Peter M. Knappeler (Berlin: Springer, 2010), 638; here van Schaik discusses the concept of culture in animals from an evolutionary perspective and why it can no longer be considered specific to humans.

23. Benett G. Galef, "Animal Culture," *Human Nature* 3, no. 2 (1992): 157–78.

24. Laland, Odling-Smee, and Feldman, "Niche Construction," 136; Carel P. van Schaik, "Animal Culture: Chimpanzee Conformity?," *Current Biology* 22, no. 10 (2012), 402.

25. Kalevi Kull, "Adaptive Evolution without Natural Selection," *Biological Journal of the Linnean Society* 112, no. 2 (2014): 287–88.

26. Tim Flannery, personal communication, 2010.

27. Peter Hiscock, *Archaeology of Ancient Australia* (London: Routledge, 2008), 52–54; Prideaux et al., "Timing and Dynamics," 22157–62; Vickers-Rich and Hewitt-Rich, *Wildlife of Gondwana*; Danielle Clode, *Prehistoric Giants: The Megafauna of Australia* (Melbourne: Museum of Victoria, 2009); Ken McNamara and Peter Murray, *Prehistoric Mammals of Western Australia* (Perth: Western Australian Museum, 2010).

28. McNamara and Murray, *Prehistoric Mammals*.

29. Christopher E. Doughty, Adam Wolf, and Christopher B. Field, "Biophysical Feedbacks between the Pleistocene Megafauna Extinction and Climate: The First Human-Induced Global Warming?," *Geophysical Research Letters* 37, no. 15 (2010).

30. Anne Forsten, "Horse Diversity through the Ages," *Biological Reviews* 64 (1989): 279–304; Vera Warmuth, Anders Eriksson, Mim Ann Bower, Graeme Barker, Elizabeth Barrett, Brian K. Hanks, Schuicheng Li, David Lomitashvili, Maria Ochir-Goryaeva, Grigory V. Vasiliy Sizonov, and Andrea Manica, "Reconstructing the Origin and Spread of Horse Domestication in the Eurasian Steppe," *Proceedings of the National Academy of Sciences* 109 (2012): 8202–6; Brian A. Hampson, Melody A. de Laat, Paul C. Mills, and Christopher C. Pollitt, "Distances Travelled by Feral Horses in 'Outback' Australia," *Equine Veterinary Journal* 42 (2010): 582–86.

31. G. Martin, "Small Native Mammals," 4–7; Mark J. Garkaklis, J. S. Bradley, and R. D. Wooller, "The Relationship between Animal Foraging and Nutrient Patchiness

in South-West Australian Woodland Soils," *Australian Journal of Soil Research* 41, no. 4 (2003): 665.

32. Martin, "Small Native Mammals," 4–7; Garkaklis, Bradley, and Wooller, "Relationship," 665; E. J. Eyre, *Journals of Expeditions into Central Australia, and Overland from Adelaide to King George's Sound, in the Years 1840–1; . . . Including an Account of the Manners and Customs of the Aborigines and the State of Their Relations with Europeans*, 2 vols. (London: T. W. Boone, 1845). Eyre recognized the Aboriginal SES, as did George Grey; George Grey, *Journals of Two Expeditions of Discovery in North-West and Western Australia, during the Years 1837, 1838 and 1839, under the Authority of Her Majesties Government, Describing Many Newly Discovered Important and Fertile Districts, with Observations on the Moral and Physical Condition of the Aboriginal Inhabitants*, vol. 2 (London: T. W. Boone, 1841).

33. J. S. Roe, "Report on an Expedition to the South-Eastward of Perth, in Western Australia, between the Months of September, 1848, and February, 1849 under the Surveyor-General Mr. John Septimus Roe," *Journal of the Royal Geographical Society of London* 22 (1849): 3.

34. Karl Dimer, *Elsewhere Fine* (Bunbury, WA: South West Printing and Publishing, 1989).

35. Geoffrey C. Bolton, *Spoils and Spoilers: A History of Australians Shaping Their Environment* (Crows Nest, NSW: Allen and Unwin, 1992), 11–12. This concept is discussed in many articles, including Cranford S. Holling, Lance H. Gunderson, and Garry D. Peterson, "Sustainability and Panarchies," in *Panarchy: Understanding Transformations in Human and Natural Systems*, edited by Lance H. Gunderson and Cranford S. Holling (Washington, DC: Island Press, 2002), 63–102.

36. Crosby, "Ecological Imperialism," 105–6.

37. Dimer, *Elsewhere Fine*.

38. Thomas G. Hungerford, *Diseases of Livestock* (Sydney, NSW: McGraw-Hill, 1975).

39. Wayne Linklater, Elissa Cameron, Kevin Stafford, and Clare Veltman, "Social and Spatial Structure and Range Use by Kaimanawa Wild Horses (*Equus caballus: Equidae*)," *New Zealand Journal of Ecology* 24, no. 2 (2000): 139–52.

40. J. S. Roe, "Report," 13–14.

41. J. P. Brooks, *Journal of a Trip to Eucla Undertaken in the Spring of 1874* (Eucla, WA: Esperance Museum, HS/832).

42. The "remount market" here refers to the breeding of young horses to sell each year to the British Army in India; Australian Light Horse Studies Centre, "Remount Section, AIF, Contents. Topic: AIF-DMC—Remounts," Saturday, 14 November 2009, http://alh-research.tripod.com/Light_Horse/index.blog?topic_id=1116006 (11 February 2017).

43. R. Ericson, *The Dempsters* (Nedlands: University of Western Australia Press, 1978), 77–78.

44. Ericson, *Dempsters*, 77–90; 252–65.

45. P. R. Bindon, "Aboriginal People and Granite Domes," *Journal of the Royal Society of Western Australia* 80 (1997): 173–79. "Gnamma" is an Aboriginal word for water-containing hole or pool of water in a rock. These may be wide shallow depressions or holes over 2 meters in depth containing thousands of liters of water. Aboriginal people

used fire to shatter the granite and augment the natural weathering and disintegration of granite to further deepen and enlarge the natural depression or hole.

46. Dimer, *Elsewhere Fine*, 42–43; Karl Dimer, *Karl Dimer—Pastoralist: Station Life from the Turn of the Century*. Transcribed from speech to Esperance Historical Society, 1973.

47. Dimer, *Elsewhere Fine*.

48. Dimer, *Elsewhere Fine*, 249.

49. Dimer, *Elsewhere Fine*, 268.

50. Dimer, *Elsewhere Fine*, 181.

51. Dimer, *Elsewhere Fine*, 268.

52. Dimer, *Elsewhere Fine*, 169.

53. Dimer, *Elsewhere Fine*, 370.

54. Patrick Duncan, "Determinants of the Use of Habitat by Horses in a Mediterranean Wetland," *Journal of Animal Ecology* 52, no. 1 (1983): 93–109.

55. Duncan, "Determinants"; http://australianmuseum.net.au/march-flies (11 February 2017).

56. Hampson et al., "Distances."

57. N. Frank, R. J. Geor, S. R. Bailey, A. E. Durham, and P. J. Johnson, "Equine Metabolic Syndrome," *Journal of Veterinary Internal Medicine* 24, no. 3 (2010): 467–75; Hampson et al., "Distances," 582–86; Christopher C. Pollitt, "Equine laminitis," *Clinical Techniques in Equine Practice* 3 (2004): 34–44. Laminitis results in the outer nail of the hoof separating from the underlying tissues with rotation of the pedal bone, and if not treated can result in permanent hoof deformity and lameness.

58. Brendan Nicholas, "Vegetation in Platt," in *Esperance Region Catchment Planning Strategy*, edited by John Platt, Brendan Nicholas, Rod Short, and Stephen Gee (Esperance, WA: Esperance Land Conservation District Committee, 1996), 11–19.

59. Linklater, Cameron, Stafford, and Veltman, "Social and Spatial Structure," 139–52.

60. Trent Ridgway, personal communication, 2013. The Ridgway family has a lease over a huge lake system for gypsum mining and sometimes see brumbies in the hazardous, hole-ridden, limestone flats with a broken leg; personal observations of horses shot and left along tracks and around water points.

61. The concept of "animal culture"' is gradually being accepted by a number of researchers as a way to explain how knowledge is passed on to group members and offspring other than through natural selection. See C. P. van Schaik, "Animal Culture," 402; Grandin and Johnson, *Animals in Translation*.

62. Jonaki Bhattacharyya and Stephen D. Murphy, "Assessing the Role of Free-Roaming Horses in a Social-Ecological System," *Environmental Management* 56 (2015): 433–46.

63. Bhattacharyya and Murphy, "Free-Roaming Horses," 441.

64. Glenn Albrecht, Clive R. McMahon, David Bowman, and Corey J. A. Bradshaw, "Convergence of Culture, Ecology, and Ethics: Management of Feral Swamp Buffalo in Northern Australia," *Journal of Agricultural and Environmental Ethics* 22 (2009): 361–78.

65. A. R. E. Sinclair, "The Role of Mammals as Ecosystem Landscapers," *ALCES* 39 (2003): 161–76; N. Thompson Hobbs, "Modification of Ecosystems by Ungulates," *Journal of Wildlife Management* 60, no. 4 (1996): 699, 702.

66. Climate Commission, *The Critical Decade: Western Australia Climate Change Impacts*, https://www.climatecouncil.org.au/uploads/e0d4e50478b96d1a50c821b7b2c022a4.pdf (12 February 2017).

67. Hobbs, "Role of Mammals," 703–4, 706.

68. Pernille J. Naundrup and Jens-Christian Svenning, "A Geographic Assessment of the Global Scope for Rewilding with Wild-Living Horses (*Equus ferus*)," *PLoS One* 10 (2015), doi: http://dx.doi.org/10.1371/journal.pone.0132359 (12 February 2017).

69. P. B. Mitchell, "Historical Perspectives on Some Vegetation and Soil Changes in Semi-Arid New South Wales," *Vegetatio* 91, no. 1–2 (1991): 169–82; and in *Vegetation and Climate Interactions in Semi-Arid Regions*, edited by A. Henderson-Sellers and A. J. Pittman (Dordrecht, Netherlands: Kluwer Academic Publishers, 1991), 175–79.

70. John F. Bruno, John J. Stachowicz, and Mark D. Bertness, "Inclusion of Facilitation into Ecological Theory," *Trends in Ecology and Evolution* 18, no. 3 (March 2003): 119–25; Stephen H. Roxburgh, Katriona Shea, and J. Bastow Wilson, "The Intermediate Disturbance Hypothesis: Patch Dynamics and Mechanisms of Species Coexistence," *Ecology* 85, no. 2 (2004): 359–71.

71. David M. Wilkinson, "The Disturbing History of Intermediate Disturbance," *Oikos* 84, no. 1 (1999): 145–47.

72. Dale G. Nimmo and Kelly K. Miller, "Ecological and Human Dimensions of Management of Feral Horses in Australia: A Review," *Wildlife Research* 34, no. 5 (2007): 408, 411.

73. Arian Wallach, "Bettongs and Bantengs: Welcome to Australia's Wild Anthropocene!," *Rewilding—Special Edition, Journal of National Parks Association of NSW* 60, no. 1 (Autumn 2016): 28–29.

74. Rule et al., "The Aftermath of Megafaunal Extinction," 1483–86.

75. Kathryn Massey and Mae Lee Sun, *Brumby: A Celebration of Australia's Wild Horses* (Wollombi, NSW: Exisle Publishing, 2013), 1–2.

Epilogue

A Global History of Species Introduction and Invasion: Reconciling Historical and Ecological Paradigms

BRETT M. BENNETT

As this book has shown, one of the defining features of European imperialism was the purposeful introduction of foreign organisms, especially plants and animals, into every country and colony in the world. Europe's nascent connection with most of the inhabited world—the Americas, Africa, and Asia—in the 1500s heralded a watershed era in global integration. During the following centuries, naturalists and governments supported efforts to classify species and move them around the world to serve human desires. Colonial ideologies justified, and sometimes required, the introduction of new species for food, fuel, shade, building materials, and more. Efforts to acclimatize new species for profit and pleasure peaked roughly between 1870 and 1939, the age of European high imperialism discussed in the previous chapters. Never before, and probably never again, will people purposefully introduce species into so many parts of the world on such a large scale. Although more total number of species are being introduced globally than ever before, the vast majority of these are unintentional invertebrate introductions rather than purposeful additions of plants or animals.[1]

Most of the new organisms that were introduced did not spread widely or cause problems, but a number did. Some of the worst consequences of these introductions—hordes of European rabbits (*Oryctolagus cuniculus*) in Australia and the like—will be familiar to most readers. The rabbit in Australia is one example of a species taken outside of its native range where it resided before human action that then became naturalized (sustains a local population) and invasive (has spread widely).[2] Naturalized and invasive species are important components of the world's ecosystems. The rate and intensity of species introductions and invasions has increased since the 1500s; a recent study suggests there is no "saturation point" yet in sight.[3] The cat is now out of the bag (or more literally the boat) in places like Australia, where it has been introduced and cannot be eradicated.

Invasion ecologists, a subset of ecologists, have focused more attention on species introductions and invasion than any other discipline, but scholars in the humanities, especially environmental history and geography, have also studied these processes for many decades. Interest in introductions and invasions emerged in the 1980s in the sciences and humanities. Alfred Crosby published his seminal work *Ecological Imperialism* in 1986, only a few years after a team of researchers working with the Scientific Committee on Problems of the Environment started the first global survey of invasive species (the results were published in 1989).[4] Since then, debates in ecology and the humanities have run parallel but rarely have scholars engaged substantially with findings from the other field. Most scholars in the sciences and history continue to conceive of introduction and invasion in starkly different terms, despite efforts to develop interdisciplinary perspectives.[5]

Scientific research on introduction and invasion assumes that invasion can be understood as an ecological process that can be induced from case studies. Finding a coherent theory for invasion has been illusive, but leading scholars have developed key concepts that inform scientific research, national legislation, and management practices.[6] At the core of invasion ecology, researchers posit that nonnative species (which can include species from within a nation but from outside an ecosystem) are more likely than indigenous (those found in a region or site) species to harm the economy, displace indigenous ecological communities and individual species, and hinder ecological services, such as water. A small but nonetheless influential cohort of scientists holds different views on the underlying ecological dynamics and the extent of the negative impacts, but they are in the minority.[7]

Historians, and humanities researchers in general, have taken a different approach to species introduction and mobility. Research in the humanities is often critical of scientific perspectives, especially when they advocate management that involves changing the access to materials or the behavior of people, especially those who are poor or have less power.[8] A sizable body of research shows how examples of efforts to control introduced and invasive species often reinforced unequal gender, power, and race relations.[9] The influence of science and technology studies, postmodernism and poststructuralism, through Bruno Latour (discussed later) and others, has made many scholars critical of scientific expertise and any explanation invoking a universal theoretical principle that transcends time and space. Most researchers are wary of theories that span centuries and continents. There are exceptions, such as Alfred Crosby, but like scientific critics of mainstream invasion ecology views, they are in the minority.

Globalization and the Anthropocene as Frameworks

This chapter suggests that a framework of globalization that both acknowledges an overall pattern of homogeneity without denying heterogeneity offers one way of reconciling aspects of scientific and humanities paradigms.[10] The benefit of a globalization framework is that it requires the least changes to the methodology of each discipline, which is an obstacle to collaboration. Although every introduction or invasion is inherently local (they happen in a finite space and within discrete contexts), the vast majority of the world's introductions and invasions are the consequence of post–1500 globalization.

Globalization is understood here as a process based on growing interconnections throughout different parts of the world driven primarily by imperialism, capitalism, and migration. The economic historian A. G. Hopkins has advanced a useful model of globalization that is based on key phases in world history.[11] One benefit of this schema is that it offers clear signposts for nonhistorians. Hopkins divides post–1500 globalization into three overlapping phases—proto-globalization (1500–1850), modern globalization (1850–1950), and postcolonial globalization. Hopkins and others posit that globalization had multiple origins, but there were people (Europeans) and regions (Europe) that had a disproportionate influence in the early phases of globalization. Europe led the first stages of globalization through its trade, colonization, and industrialization, but in the first centuries its expansion was to a large extent built on older, existing trade networks in Asia, Africa, and to a lesser extent the Americas.

Studies of introduced and invasive species posit slightly different periodizations and geographies than Hopkins because their focus is on biotic globalization specifically. Alfred Crosby argued that there was a biological expansion of Europe from 900 to 1900, leading to colonization of places resembling Europe in its climate.[12] John McNeill somewhat modified Crosby's earlier dating by arguing that before 1500, Europe received more new species imports than it exported, although he agreed that by 1900, Europe once again imported rather than exported species.[13] In an influential paper on invasive species, the ecologist Phillip Hulme traced a first phase of European expansion from 1500 to 1800; a second phase of increased invasions in Europe and North America based on the industrial revolution from around the 1850s; and a third phase, the era of globalization of the past three decades or so.[14]

Scholars studying the Anthropocene have also focused on globalization. Many of the key dates of globalization and the Anthropocene align broadly. Some scholars have dated the origins of the Anthropocene to the 1600s or

1800s, a dating consistent with work on globalization.[15] Yet other scholars push the date forward and backward: some go as far back in time as the late Pleistocene, while others do not see it as beginning until the first atomic bomb was dropped.[16] In terms of introduced and invasive species, it is clear that a key shift occurred somewhere between 1600 and 1900, rather than before or after. This chapter supports a more middle rather than earlier or later dating to the Anthropocene.

In the light of these findings, it is perhaps easiest and best to conceive of two stages in ecological globalization since the 1500s based on the rate and flow of introduced species. The first period, which dates roughly from 1500 to 1900, was shaped disproportionately by European imperialism and European-led long-distance trade. These processes caused a rapid increase in the diffusion of biota from different parts of the world. During this period there was a general increase in the number of species that were introduced and became established in every region of the world, but Eurasian flora and fauna disproportionately constituted the introduced and invasive species in nontropical areas as a result of settler colonialism.

The second phase, which started around 1850–1900 and continues today, has been characterized by the increase in invasions globally and a "catching-up" of invasions in Europe and tropical regions compared to other parts of the world. This dating corresponds with the beginning of Hopkins's "modern globalization" phase. Hulme is right to point out the growing velocity of introductions from the second half of the twentieth century, but we should not forget that the late nineteenth and early twentieth centuries were periods of intense international trade; the late 1800s were only surpassed in terms of percentage of a country's trade in the 1960s.[17] Also, purposeful (rather than unintentional) introductions peaked in the late nineteenth and early twentieth centuries. Recent studies of invasive species in Europe point to 1900 as the beginning of a shift toward increased invasions in Europe, a region that previously lagged behind the temperate regions of the Americas and Australasia in terms of the total number and ecological impact of invasions.[18]

These are *rough* dates indicative of major global changes, and the dating may be earlier or later than changes for a specific place. Yet the trend is clear: there seems to be no "saturation" in the rate of introductions and the discovery of new invasions, something that suggests the process will continue to lead to a sort of ecological homogenization of many parts of the world.[19] At the same time, as I discuss later, this overall homogenization must be understood as the outcome of the world's great biological heterogeneity, which is based on distinct evolutionary histories of regions and particular taxa. Human activity and underlying ecological conditions have created heterogeneity—such as why rubber plantations developed in one country but not in another that could feasibly support them—so this model does not deny variation but nests it within a global context *tending* toward homogeneity.[20]

This periodization revives key parts of Alfred Crosby's ecological imperialism thesis, which has received substantial criticism from many historians but has been more positively appraised in invasion ecology. Crosby argued that a combination of historical, geographical, and evolutionary factors allowed Europeans, and flora and fauna adapted to regions with temperate climates similar to Europe, to successfully colonize temperate regions of the Americas and Australasia. His views have often been interpreted as environmental determinism, when in fact he gave considerable attention to socioecological interactions through his concept of the "portmanteau biota," an idea that has been used by other scholars and is largely justified based on extensive findings across a range of fields. Though many historians have challenged Crosby's methodology and evidence, a number have continued to build on his work while also modifying his ideas.[21]

Crosby's ideas on ecological change are highly relevant to understanding the past and present because they allow us to understand complex global change through a socioecological framework incorporating humans and nonhumans as actors. The geographer Paul Robbins drew explicitly from Crosby's portmanteau biota framework in 2004 to analyze invasions at smaller spatial scales.[22] Undoubtedly, some of Crosby's views and periodizations should be revised based on recent historical research, but that does not mean his overall argument about introduced plants and animals is wrong. (There is an extensive debate about his epidemiological theory of disease, which falls outside of the scope of this epilogue). In his global study of guano, the world's first widely used nitrogen-based fertilizer, Greg Cushman argues that Crosby's thesis is useful to explain the first phase of settler colonialism but it becomes less relevant after there was an emergence of a fuel- and fertilizer-intensive industrial economy in the mid-1800s.[23] The mid-1800s also corresponded to key historical changes, such as acclimatization and state science, so the slightly earlier dating (i.e., 1850s to 1870s rather than 1900) is in some ways better from a historical point of view because this is when many currently invasive species were introduced and began to spread.[24]

Somewhere between 1850 and 1900, the dynamics of globalization changed. The number of introduced species and those that became invasive became more equalized throughout the whole of the world, and especially in those regions connected by trade.[25] The rate of introduction has continued to expand and equalize throughout the twentieth century.[26] Regions such as Europe or the tropics that lagged in terms of recognized invasions are catching up to places such as Australia or North America, which experienced earlier and more extensive invasions.

Alfred Crosby's Portmanteau Biota

Alfred Crosby's 1986 book *Ecological Imperialism* is a useful entry point into this topic because it remains relevant today and is widely known by scholars in a variety of fields. Even critics are aware of its overall thesis, so it is perhaps the best way to communicate to scholars in multiple fields. For more than thirty years, Crosby's book on ecological imperialism has remained a central reference point of historical research examining the ecological impact of European colonialism. Most historians expressed concerns about Crosby's ideas rather than using them as a theory, but there has been a small but steady number of scholars who have extended his ideas. Scientists have shown far more sympathy for Crosby's arguments.

Crosby was fascinated by how facets of European ecological and agricultural conditions were adapted in the temperate climate regions of the Americas and Australasia colonized by European settlers. He described these regions as neo-Europes because they had to a significant degree been modified through colonization to resemble Europe in many facets, including landscapes and ecologies. His book did not examine tropical regions and Africa, the Middle East, or Asia because Europeans never successfully replicated, in large numbers, European ecosystems and settlements in these places, with the exception of South Africa, a region he did not deal with in the book.

Crosby argued that there was an imbalance in plants exported from Europe outward that could not be explained merely by the fact that people brought more plants from that region compared to another. He did not give statistical evidence for this claim. Crosby pointed out that many Eurasian species seemed to succeed even in the absence of European settlers, whereas this seems to have happened very rarely with reverse flows. Why did this happen? Crosby wrote: "There is little or nothing intrinsically superior about Old World organisms compared with those of the Neo-Europes." What differed was the evolutionary history of Old World organisms and the unique way in which so many plants were introduced into a region that had been transformed by colonists.

Crosby argued that a synergistic interaction between introduced species, including humans, was the primary explanation for the dominance of

Eurasian species abroad and the relatively moderate impacts Europe experienced with introductions from the regions it colonized. Crosby argued that the evolutionary history of Eurasian species gave them unique advantages in highly human modified environments, such as areas of intense agriculture or grazing. Crosby wrote, "The co-evolution of Old World weeds and Old World grazers gave to the former a special advantage after the two spread in the Neo-Europes, and added on top of that was the advantage these plants had of having evolved along with the development of Old World agriculture."[27] Mammals succeeded for somewhat different reasons: a significant proportion of mammals thrived in new environments, such as Australia or New Zealand, because they lacked large predators, which is an argument explained to some degree by biogeography.

Crosby attributed the success rate of the portmanteau biota only partly to their evolutionary strategy (e.g., a weed growing among wheat). The sheer amount of species that were introduced around the same time shocked precolonial socioecological systems. Europeans came with what Crosby called the "portmanteau biota," a plethora of organisms that had evolved with Europeans. He wrote, "The success of portmanteau biota and its dominant member, the European human, was a team effort by organisms that had evolved in conflict and cooperation over a long time."[28] The portmanteau biota included organisms with common evolutionary histories based on cohabituating in human-disturbed landscapes. These species interacted with each other in newly colonized environments in a way that, alongside human disturbance, promoted an increase in the number and rate of invasions. He described portmanteau biota as Europeans' "fellow life forms, their extended family of plants, animals and microlife—descendants, most of them, of organisms that humans had first domesticated or that had first adapted to living with humans in the hearthlands of Old World civilization. . . . Where it 'worked,' where enough of its members prospered and propagated to create versions of Europe, however incomplete and distorted, Europeans themselves prospered and propagated."[29]

It is not entirely easy to separate out the various strands of causality of Crosby's arguments, something that has led to numerous somewhat misled critiques of his work, but a simple breakdown might look something like this: Crosby gave relatively equal weight to the traits of individual species that evolved in human-modified ecosystems, to geography, to the synergistic effect caused by the introduction of so many organisms into a single place at the same time, and to human action, through disturbance, colonization, and introduction. He gave broadly equal weight to evolution, ecology, biology, geography, and also human culture and history. The dramatic and popular style of his writing—the same reason his book was a success—sometimes limited its analytic clarity. Yet his writing was clear enough to inspire debate that continues to this day.

Historical Reception of Ecological Imperialism

Crosby's book has polarized readers since its debut. Although several historians have engaged positively with his ecological imperialism thesis, the majority of historians have been cautious or openly critical of his ideas. This section focuses on criticisms of Crosby's work before discussing his generally positive reception in ecology.

Crosby's thesis is controversial because he used evolutionary, ecological, and epidemiological theories to explain European colonization in the Americas and Australasia.[30] Peter Coates chided Crosby for "coming close to reviving turn-of-the-century environmental determinism" (a reference to racialist thinking of geographers such as Ellsworth Huntington) and for denying European "culpability" in the destruction of indigenous ecosystems, populations, and lifestyles.[31] Using biology and evolution in historical explanations is viewed by many historians as dangerous for various reasons, not least because environmental determinism was a key part of pre–World War II racial theories.

Even more historians have expressed alarm that Crosby supposedly downplayed the human role in ecological and demographic change. Not long after the publication of Crosby's book, B. R. Tomlinson wrote an important rebuttal arguing that, "Much of the apparently 'natural' expansion of the plants and pathogens that occurred in the neo-Europes was closely linked to the process of human colonization and conquest."[32] Downplaying human agency is seen as justifying settler colonial violence and dispossession. It is often forgotten that Crosby wanted to downplay the idea that Europeans were superior as the primary explanation for why Europe rather than say, China, became the dominant region in the world.

Another constant criticism is that Crosby overemphasized the dominance of Eurasian species in the Americas and Australasia. Historical examples from a range of places—New Zealand, India, South Africa, the United States—demonstrate that there were many counterflows of crops (like potatoes and maize) as well as naturalized and invasive species that came from the so-called Neo-Europes.[33] James Beattie argues that New Zealand was just as much the "empire of the rhododendron" because of the presence of Asian plants, as it was of the "empire of the dandelion," which came from Eurasia.[34] Europe also received many imports from the Neo-Europes. Peter Coates writes that the "most controversial immigrant fauna in Britain are all deliberate introductions from North America: gray squirrel, mink, muskrat, signal crayfish and ruddy duck."[35] Historians of South Africa, a region of European settlement that Crosby did not focus on, argue that a large proportion of introduced species came from the Americas or Australasia.[36] Experts on India point to a similar fact.[37] Some have gone so far to suggest that Crosby's evidence of a Eurasian dominance in temperate climates is factually incorrect. William Beinart and Karen Middleton write, "in the period from 1500–1900, plant transfers may have been more evenly balanced than Crosby suggests."[38] Beinart and Middleton give no statistical breakdown for this claim. In fact, humanities scholars criticizing Crosby do not use statistics or findings from national field surveys.

Scientific Reception

Ecologists tend to have a more favorable view of Crosby. Moreover, there is a large body of research that was done without reference to Crosby (some invasion biologists know of Crosby, but many do not) that supports his ideas. Crosby's views are compatible with key tenets of contemporary understandings of invasion dynamics based on three decades' worth of empirical and theoretical research.

Just after the publication of *Ecological Imperialism*, the invasion biologist Daniel Simberlof wrote, "This is a provocative, novel hypothesis. It will be interesting to see how it fares when subjected to rigorous testing."[39] Francesco Di Castri, an Italian-born Chilean expert on invasion who helped direct the first global survey of invasive species by SCOPE in the 1980s, came to a similar conclusion as Crosby. Di Castri argued that Eurasian plants, and especially those from the European Mediterranean, seemed to be more invasive in New World habitats with Mediterranean climates than New World plants were invasive in the European Mediterranean.[40] Di Castri agreed explicitly with Crosby on this point.[41]

Crosby's point about asymmetry has been borne out by extensive surveys of invasive species.[42] Even today, a disproportionate amount of naturalized and invasive species in the Americas and Australasia originate from Eurasia compared to other parts of the world. In 2013, Simberloff wrote that there is "no doubting the general pattern" of Old World species being disproportionately invasive in these regions.[43] This dominance is particularly apparent in fields and agrarian environments rather than in closed forests and deserts.[44] This asymmetry occurs globally, including in warmer Mediterranean climates as well as temperate climates, so Crosby's theory can be expanded somewhat beyond his original intention.

Various explanations have been offered to explain this asymmetry. There is undoubtedly a strong historical component that reflects contingent events (e.g., one plant was brought to one place but not another; farmers or the state had greater capital to import more species and support them). Obviously, European migrants were, at least initially, more likely to bring species from Eurasia that they relied upon for food, materials, and aesthetics rather than from regions they did not know. The growth of global trade around the mid-1800s unleashed a much greater number of potential invasive species on the world, something that is recorded in surveys of invasions. According to the theory of propagule pressure, the larger the population, the more likely it is to become naturalized and then invasive.

Though necessary, the historical argument—that Europeans brought what they had on hand and the sheer number of them means that more became invasive—may not be sufficient. Simberloff weighted the competing evidence and claims to suggest that although history influenced this asymmetry, it alone did not explain the imbalance: "In sum, the sites and sources of invasions can often be explained by the history of human activities that move species. However, an apparent propensity of Eurasian species to invade elsewhere cannot wholly be attributed to history."[45] This is to say, there were unique socioecological dynamics associated with European colonization and there may be underlying biogeographic and evolutionary reasons as well.[46] Two processes—invasional meltdown and the evolutionary history of Eurasian plants—have been invoked to account for this imbalance.

In an important article in the first issue of the journal *Invasion Biology,* David Simberloff and Betsy Von Holle build on Crosby directly by suggesting that invasion may be enhanced when there is an "invasional meltdown," the establishment of multiple species populations that directly or indirectly facilitate a new invasion.[47] Their argument challenged scholars who saw biotic resistance, the fitness of an ecosystem to resist invaders, as the primary way to measure the potential invasability of an area. Instead of focusing primarily on the resilience of an ecosystem, they posited that researchers need to study how introduced species created unique ecological dynamics that could synergistically work to facilitate invasion. After citing Crosby's theory as one alternative model to biotic resistance because it emphasizes interspecies interactions among introduced species, they go on to offer a definition that is heavily indebted to Crosby's portmanteau biota: "We suggest the term 'invasional meltdown' for the process by which a group of nonindigenous species facilitate one another's invasion in various ways, increasing the likelihood of survival and/or of ecological impact, and possibly the magnitude of impact."[48] Their article, which has been cited extensively and is one of the key concepts created in the field of invasion biology, had the intended effect of turning attention toward the mutualistic dynamic of imported species.[49]

Crosby's other ideas, such as the evolutionary changes associated with agriculture and cities, have been explored by scientists elsewhere. Research agrees with Crosby that the coevolution of weeds and pests may help explain the success of feral plants and animals in agrarian environments.[50] Unique evolutionary conditions can be found outside of Eurasia. Reviewing global grass invasions, scholars have argued that Southern Africa is a key producer of the world's invasive grasses (likely due to fire and other pressures), much like Australian trees (adapted to drought and fire and respond well to high levels of nutrients) contribute a disproportionate percentage of the world's invasive trees.[51] Asia produces a significant percentage of the invasive marine species in ports.[52] These geographic and evolutionary factors must be a factor in how we analyze historical ecological change.

Crosby's 1900 dating holds up remarkably well. Recent research indicates that Europe has been experiencing a growing rate of invasions since 1900. A European Union report issued in 2016 noted, "Results showed that the number of alien species in Europe has increased linearly over the past 100 years, leading to a fourfold increase in numbers since 1900."[53]

The increase of total introductions and invasions has been explained using various theories. Some invasion researchers argue that species can "lag," that is, reside in a place for a long time before spreading. Some ecologists argue for an "invasion debt" theory which proposes that the potential debt of invaders can be calculated by the total number of species in a place, their residency time, traits, and evolutionary background.[54] There is a correlation between economic activity and introductions. Europe was in some ways a prime target for getting its ecological comeuppance because it had the world's most extensive trade links.

This economic view reinforces the pioneering ideas of one of the "founders" of invasion ecology, Charles Elton, who saw invasions as a global phenomenon linked closely to increased trade and migration.[55] Elton was focused on gleaning universal scientific concepts from "biological explosions" of invasive species rather than historicizing specific periods in the history of invasions.[56] Elton's belief in the cosmopolitan nature of invasions is confirmed by historical and economic research that shows an intensification of global trade and mobility during key periods (i.e., 1870 to 1914 and 1970s–present) from the late nineteenth century on. Invasion ecologists have shown that every part of the world has a greater number of naturalized and invasive species than ever before in known history.[57] There is a general process of ecological homogenization. Most of the worst invasive species in one country are considered equally bad in another part of the world. Research in ecology suggests that there are key underlying geographic and evolutionary factors shaping invasion that are then influenced by historical contingency.

Having a Critical and Constructive Engagement with Science

The tale of two Crosbys—a more critical reception in history and a more positive reception in science—raises important questions. Why did Crosby receive a more positive reception in the sciences than in the humanities? Did scholars in the humanities not know about developments in invasion ecology or did they just not think them to be relevant? And most important, how can this lesson inform future engagements between history and ecology?

The lack of engagement with scientific research on introduction and invasion reflects a wider distancing of the humanities and sciences since around the time Crosby wrote his book. Jo Guldi and David Armitage argue in *The History Manifesto* that, since the 1980s, the profession of history has suffered increasing neglect and a decline of public importance because the discipline pursued approaches that were often openly antagonistic to overtly materialist, empirical, or scientific viewpoints.[58] Bruno Latour, Michel Foucault, and other key thinkers called on scholars to deconstruct positivism, scientism, and materialism inherited from the Enlightenment and Scientific Revolution. Scholars advocated theoretical positions that undermined traditional causality and popular scientific conceptions such as objectivity, facts, or truth.

For various reasons, it is increasingly difficult to sustain a model of history that rejects insights from the sciences and separates itself from questions of materiality. Yet how historians approach ecology is something that is being discovered on an individual-by-individual basis. Science (including ecology) is not usually taught in history PhD programs so historians, unless they take another degree, will need to be somewhat self-trained. The environmental historian William Beinart rightly admitted in his study of South

African conservation that it "is very difficult for a historian untrained in science to evaluate . . . [scientific] studies."[59] He nonetheless concluded somewhat more optimistically: "but it is possible to discuss them in light of other evidence" that historians can know.[60]

The stumbling blocks to engaging with science are real, but they need not stop scholars from at least reading in other fields and putting science in a conversation with history, as Beinart did. Scholars in a variety of fields utilize knowledge from areas in which they are not experts.[61] For instance, an ecologist who works on a multiauthored paper may not be familiar with the literature cited in the same paper by a mathematician or geologist. The team-based approach is foreign to most historians, who are locked into a single author and book-based model for tenure and promotion (at least that is the ideal for leading North American institutions, which have a disproportionate influence on scholarship).

In lieu of further formal study, historians can train themselves to learn the broad outlines of key debates and talk with experts outside of their area of specialty. Ecological concepts, at least in their more generalized forms, are not too difficult to comprehend. The terminology can be learned and experts who hold different opinions can be consulted. In my personal experience, ecologists are open to engagement and want to work with humanities scholars. The work of influential historians—Libby Robin, Jane Carruthers, and Simon Pooley, among others—shows that one can work with scientists without being a scientist.[62]

Historians often shy away from interdisciplinary engagements because it requires an epistemic shift away from one mode of evidence and reasoning to another one.[63] Scientists (ecologists in this case) and historians use generalization and causality differently. Recognizing these differences is the first step to overcoming them to some degree. The historian John Lewis Gaddis suggests in his book on the writing of history that historians use "particular generalizations" from discrete events that are contextualized within their specific time and place.[64] Scientists, he argues, use a different method that relies more on "universal generalizations" and theoretical models formed from examples. An ecologist studying invasion processes today is more likely to generalize present examples in understanding changes in the past, whereas a historian would suggest that each period of history must be taken on its own terms.

A wariness about making generalizations now characterizes much of the scholarship in the fields of global, imperial, and transnational history, where species introduction and invasion have most frequently been discussed.[65]

Many scholars in these fields have shifted toward using networked and other decentralized approaches to tracing mobility and flows to overcome older power analyses using concepts such as core and periphery.[66] For instance, historians who criticize Crosby have offered no other systematic global model and thus they posit (explicitly or implicitly) that randomness and historical contingency characterized the dynamics of the era.[67] Examples of counter-flows are used to show that biotic exchange happened in other parts of the world. This is a point that Crosby understood perfectly well. Single case studies, discussed earlier, often focus on places (South Africa or India) or examples (noninvasive species requiring significant human support, such as *Eucalyptus*) well outside of the geographic limitations of Crosby's thesis and thus cannot credibly be seen as a criticism of his specific claims. In all these cases, the exception does not actually prove the rule.

Using ecology, biogeography, or evolution to provide causal explanations for historical change runs contrary to how many (but not all) historians assign causality and agency in history.[68] Prominent scholars of imperial and global history, such as Alan Lester, have argued that historians should not or cannot assign causation in history.[69] In this view, causation—the discovery of putative origins of processes—is associated with an older mode of historical inquiry that focused on Europe as the origins of global capitalism, imperialism, democracy, and industrialization. In many ways, Crosby fits this older model: he sought to explain the dominance of Europeans in certain climates and environments using a similar "origins" structure. Ironically, Crosby sought to downplay European intellectual or cultural superiority by highlighting nonhuman drivers of Eurasia (including Asia), but his point has been lost over the decades. The position of this chapter is that historians need to at least engage with questions of causality in other fields and try to align with some of our thinking. There are significant problems associated with causality, but rejecting it entirely leaves us in a worse position. If we reject causality, even in its broad forms, we run the risk of seeing the world as contingent, a perspective that in some ways undermines the value of history as a didactic tool.

As demonstrated by chapters in this book, recent developments in the humanities—such as animal studies and posthuman thinking—have raised the prospect of assigning agency to nonhumans. Yet in many respects, this potential has been muted due to the lingering influence of certain strains of science and technology studies and poststructuralist, and postmodern thinking. Key ideas include: the asymmetry thesis (which does not judge the validity of historical claims), Latour's actor-network theory (which denied the ability to use theory to assign causal priority to agents/objects that interact in what Latour described as a "network"), and constructivism (which posited that ideas are constructs not based on fundamental physical reality). Actor-network theory (ANT) has been widely used to trace human-nature entanglements in a way that does not privilege human agency.[70] Not all scholars take these theories to their logical conclusion, but the influence of these intellectual strains has shaped an emergent field of environmental humanities that is often highly critical of the politics of ecology (as evidenced in the term "invasive species").

There has been a pushback against the excesses of these views. Bruno Latour, who more than anyone shaped a critical stance toward scientific knowledge, wrote in a critique in 2004:

> Entire Ph.D. programs are still running to make sure that good American kids are learning the hard way that facts are made up, that there is no such thing as natural, unmediated, unbiased access to truth, that we are always prisoners of language, that we always speak from a particular standpoint, and so on, while dangerous extremists are using the very same argument of social construction to destroy hard-won evidence that could save our lives. Was I wrong to participate in the invention of this field known as science studies? Is it enough to say that we did not really mean what we said?[71]

Latour did not reject everything he wrote about, but he rightly suggested that theoretical lenses are tools for understanding a complex world, not ontological realities unto themselves. Despite Latour's plea, some fields (e.g., history of science and environmental humanities) influenced by his work from the 1980s and 1990s adhere somewhat closer to his earlier rather than contemporary views on scientific knowledge.

In a similar vein, Guldi and Armitage argue that the lack of engagement with the sciences and social sciences is problematic because it has meant that historians have less relevance in public debates. Even though there are pitfalls in trying to trace patterns across long periods of time, they argue that historians are well equipped to do this because they are aware of the importance of context. None of these scholars, from Guldi to Latour, advocates using science as "truth," but they suggest that we must engage seriously with scientific "matters of concern," such as global warming and invasive species, raised by the scientific community. This chapter echoes these sentiments and hopes to encourage more historians to think both *constructively* as well as critically about science.

How do we all work together when the humanities and sciences view generalization, causation, and evidence differently? We can turn both to pragmatism and constructivism, traditions that urge us to recognize that all disciplinary knowledge is a particular lens onto the world rather than the world itself. The production of discipline-specific knowledge should be a means to an end product that must be negotiated with other disciplines and social groups to enact ideas into policy and action. History should not only be read by historians. Disciplines are not useful if they keep us from engaging with other disciplines or the public. By no means does this mean that historians should quit trying to contextualize periods of history. Far from it. Guldi and Armitage rightly point out these unique contextual skills and research abilities—skills based on a focus on particulars as well as generalizations—are required when assessing long-term patterns. But doing good interdisciplinary work is going to require many humanists to at least engage constructively, not only critically, with scientific concepts and scientists.

Historians should write research for their peers but also engage with other disciplines using the concepts in those fields while fully recognizing their limitations. Scientists and many social scientists increasingly conceive of a landscape in terms that capture its range of influences, from human action all the way down to basic biophysical processes. Historians generally shy away from making universal generalizations, but we should not be afraid to learn about the evidence and theories of other fields in order to create hypothetical historical scenarios. For instance, research on the population dynamics of current plant invasions and genetic analysis of species (which can tell what region of a foreign country they come from and even *when*) can be used to think about historical events where we have no written, oral, or visual records. The "social" in socioecological recognizes that natural systems are inherently interlinked intellectually and materially with humans. This conceptual tool offers historians the space to engage directly with other scholars. There are legitimate critiques of this approach, but it would be a shame if we let theoretical purity stop historians from working with scientists directly on issues of common concern.

Conclusion

Scholars seeking to understand the history, patterns, and possible futures of biotic globalization should engage more with findings from history, ecology, and a variety of other disciplines (e.g., geography) when developing macroscale historical interpretations. A large body of ecological research suggests that there were unique ecological patterns associated with the expansion of European settler colonies. The dating of 1900, first put forward by Crosby, has been confirmed subsequently in ecology and history as a useful period to mark the shift from one phase of ecological globalization to another. That ecologists find consistent patterns of convergent ecological changes throughout the world suggests that biologic globalization as a process was based solely on human agency. It is not sufficient to discredit global patterns with single case studies of microhistory. The challenge for those who disagree with Crosby or science-informed interpretations is to offer an alternative hypothesis that engages directly (even if critically) with the argument and the evidence discussed and put forth in this chapter.

Notes

1. Phillip Hulme, "Trade, Transport and Trouble: Managing Invasive Species Pathways in an Era of Globalization," *Journal of Applied Ecology* 46, no. 1 (2009): 10–18.

2. For definitions, see David M. Richardson, Petr Pyšek, Marcel Rejmánek, Michael G. Barbour, F. Dane Panetta, and Carol J. West, "Naturalization and Invasion of Alien Plants: Concepts and Definitions," *Divers Distribution* 6 (2000): 93–107. In this epilogue I use the terms "naturalization" to denote a species that has been introduced and has a population within a new region and "invasion" to refer to self-reproducing species that colonize in areas where there is no historical record of their occupation. Many historians may prefer to use other terms, but these terms are useful for their technical description (i.e., something that is introduced, something that establishes itself, and something that reproduces) irrespective of impact.

3. Richard N. Mack, Daniel Simberloff, W. M. Lonsdale, Harry Evans, Michael Clout, and Fakhri A. Bazzaz, "Biotic Invasions: Causes, Epidemiology, Global Consequences, and Control," *Ecological Applications* 10, no. 3 (June 2000): 689–710. For a historian's view on these challenges, see William Beinart, "Bio-invasions, Biodiversity, and Biocultural Diversity: Some Problems with These Concepts for Historians," *RCC Perspectives* 1 (2014): 75–80, 75.

4. James A. Drake, Harold A. Mooney, F. Di Castri, R. H. Groves, F. J. Kruger, M. Rejmánek, and M. Williamson, eds., *Biological Invasions: A Global Perspective* (Chichester, UK: Wiley, 1989).

5. Ana S. Vaz, Christoph Kueffer, Christian A. Kull, David M. Richardson, Stefan Schindler, A. J. Muñoz-Pajares, Joana R. Vicente, João Martins, Cang Hui, Ingolf Kühn, and João P. Honrado, "The Progress of Interdisciplinarity in Invasion Science," *Ambio-stockholm-46* 4 (2017). Also see Cang Hui and David Richardson, *Invasion Dynamics* (Oxford: Oxford University Press, 2017), chap. 1; David Richardson, Jane Carruthers, Cang Hui, Fiona A. C. Impson, Joseph T. Miller, Mark P. Robertson, Mathieu Rouget, Johannes J. Le Roux, and John R. U. Wilson, "Human-Mediated Intro-

ductions of Australian Acacias—A Global Experiment in Biogeography," *Diversity and Distributions* 17, no. 5 (2011): 771–87.

6. For the most up-to-date survey, see Richardson and Hui, *Invasion Dynamics*.

7. See one critical opinion: Mark A. Davis, "Don't Judge Species on Their Origins," *Nature International Journal of Science* 474 (July 2011), 153–54, http://www.nature.com/nature/journal/v475/n7354/full/475036a.html (30 April 2019).

8. The list is not exhaustive, but representative articles from different fields include: Susanna Lidström, Simon West, Tania Katzschner, M. Isabel Pérez-Ramos, and Hedley Twidle, "Invasive Narratives and the Inverse of Slow Violence: Alien Species in Science and Society," *Environmental Humanities* 7, no. 1 (2016): 1–40; Jonathan Clark, "Uncharismatic Invasives," *Environmental Humanities* 6, no. 1 (2015): 29–52; Charles Warren, "Perspectives on the 'Alien' versus 'Native' Species Debate: A Critique of Concepts, Language and Practice," *Progress in Human Geography* 31, no. 4 (2007): 427–46; Jean Comaroff and John L. Comaroff, "Naturing the Nation: Aliens, Apocalypse and the Postcolonial State," *Journal of Southern African Studies* 27, no. 3 (2001): 627–51.

9. This is a representative but not exhaustive list of studies. See Lance van Sittert, "'Our Irrepressible Fellow-Colonist': The Biological Invasion of Prickly Pear (opuntia Ficus-Indica) in the Eastern Cape C. 1890–C. 1910," *Journal of Historical Geography* 28, no. 3 (2002): 397–419; William Beinart and Luvuyo Wotshela, *Prickly Pear: The Social History of a Plant in the Eastern Cape* (Johannesburg: Wits University Press, 2011); Jodi Frawley and Iain McCalman, "Invasion Ecologies: The Nature/Culture Challenge," in *Rethinking Invasion Ecologies from the Environmental Humanities*, edited by Jodi Frawley and Iain McCalman (Oxford: Routledge, 2014), 3–14; Libby Robin and Jane Carruthers, "National Identity and International Science: The Case of *Acacia*," *Historical Records of Australian Science* 2, no. 1 (2012): 34–54; Simon Pooley, "Pressed Flowers: Notions of Indigenous and Alien Vegetation in South Africa's Western Cape, c. 1902–1945," *Journal of Southern African Studies* 36, no. 3 (2010): 599–618; Jane Carruthers, Libby Robin, Johan Hattingh, Christian Kull, Haripriya Rangan, and Brian W. van Wilgen, "A Native at Home and Abroad: The History, Politics, Ethics and Aesthetics of *Acacia*," *Diversity and Distributions* 17, no. 5 (2011): 810–21; Natasha Nongbri, "Plants Out of Place: The 'Noxious Weeds' Eradication Campaign in Colonial South India," *Indian Economic & Social History Review* 53, no. 3 (2016): 343–69; Peter Coates, *American Perceptions of Immigrant and Invasive Species: Strangers on the Land* (Berkeley: University of California Press, 2006).

10. Mark Davis used the term in 2003. This article does not advance or criticize Davis and uses the term neutrally. Mark A. Davis, "Biotic Globalization: Does Competition from Introduced Species Threaten Biodiversity?," *Bioscience* 53, no. 5 (May 2003).

11. See A. G. Hopkins, ed., *Globalisation in World History* (London: Pimlico, 2002).

12. Alfred W. Crosby, *Ecological Imperialism: The Biological Expansion of Europe, 900–1900* (Cambridge: Cambridge University Press, 1986; 2nd ed. 2004).

13. J. R. McNeill, "Europe's Place in the Global History of Biological Exchange," *Landscape Research* 28, no. 1 (2003): 33–39.

14. Hulme, "Trade, Transport and Trouble," 10–18.

15. For 1600s, see Lewis and Maslin. Also see Wolfgang Lucht, "The Human Planet: How We Created the Anthropocene," *Nature Journal of Science* 558 (June 2018): 26–27, https://www.nature.com/articles/d41586-018-05315-6 (30 April 2019). For the 1800s, see W. Steffen, J. Grinevald, P. Crutzen, and J. McNeill, "The Anthropocene: Conceptual and Historical Perspectives," *Philosophical Transactions of the Royal Society A: Mathematical, Physical and Engineering Sciences* 369, 1938 (2011): 842–67.

16. Jan Zalasiewicz, Mark Williams, Alan Haywood, and Michael Ellis, "The Anthropocene: A New Epoch of Geological Time?," *Philosophical Transactions of the Royal Society A: Mathematical, Physical and Engineering Sciences* 369 (2011): 835–41.

17. World Trade Organization, *World Trade Report 2013* (Lanham, MD: Bernan Press, 2013), 47.

18. F. Essl, S. Dullinger, W. Rabitsch, P. E. Hulme, K. Hülber, V. Jarošík, and I. Kleinbauer, "Socioeconomic Legacy Yields an Invasion Debt," *Proceedings of the National Academy of Sciences of the United States of America* 108, no. 1 (2010).

19. See F. Essl et al., "Socioeconomic Legacy," 203–7. For further development, see Rouget Mathieu et al., "Invasion Debt—Quantifying Future Biological Invasions," *Diversity and Distributions* 22, no. 4 (2016): 445–56.

20. Corey Ross, "Developing the Rain Forest: Rubber, Environment and Economy in Southeast Asia," in *Economic Development and Environmental History in the Anthropocene: Perspectives on Asia and Africa*, edited by Gareth Austin (London: Bloomsbury Academic, 2017), 199–218.

21. Notable examples include Elinor Melville, *A Plague of Sheep: Environmental Consequences of the Conquest of Mexico* (Cambridge: Cambridge University Press, 1994) and Greg Cushman, *Guano and the Opening of the Pacific World: A Global Ecological History* (Cambridge: Cambridge University Press, 2013), xvi.

22. This global-scale focus builds on and agrees with Paul Robbins's analysis of how the concept of "portmanteau biota" combines natural and cultural dynamics into a coherent model. See Paul Robbins, "Comparing Invasive Networks: Cultural and Political Biographies of Invasive Species," *Geographical Review* 94, no. 2 (2004): 139–56, 143.

23. Cushman, *Guano and the Opening of the Pacific World*, 17–18. See his comments on Crosby on 71, 110.

24. This is a key point in F. Essl, "Socioeconomic Legacy Yields an Invasion Debt."

25. For a discussion of the intensity of economic integration and human migration in the late nineteenth and early twentieth centuries, see "Introduction" in Michael Bordo, Alan M. Taylor, and Jeffrey G. Williamson, *Globalization in Historical Perspective* (Chicago: University of Chicago Press, 2007). See the threefold stages of A. G. Hopkins, "Is Globalization Yesterday's News," *Itinerario* 41, no. 1 (2017): 109–28.

26. Hulme, "Trade, Transport and Trouble," 11, fig. 1.

27. Hulme, "Trade, Transport and Trouble," 11, fig. 1.

28. Hulme, "Trade, Transport and Trouble," 293.

29. Hulme, "Trade, Transport and Trouble," 89.

30. See the discussion by Dipesh Chakrabarty, "The Climate of History: Four Theses," *Critical Inquiry* 35, no. 2 (Winter 2009): 197–222, 214–15.

31. Peter Coates, *Nature: Western Attitudes since Ancient Times* (Oxford: Blackwell Press, 1998), 102.

32. B. R. Tomlinson, "Empire of the Dandelion: Ecological Imperialism and Economic Expansion, 1860–1914," *Journal of Imperial and Commonwealth History* 26, no. 2 (1998): 84–99, 90.

33. Tomlinson, "Empire of the Dandelion," 90.

34. James Beattie, "'The Empire of the Rhododendron': Reorienting New Zealand Garden History," in *Making a New Land: Environmental Histories of New Zealand,* edited by Tom Brooking and Eric Pawson (Dunedin, New Zealand: Otago University Press, 2013), 241–57.

35. Peter Coates, *American Perceptions of Immigrant and Invasive Species: Strangers on the Land* (Berkeley: University of California Press, 2006), 24.

36. Lance van Sittert, "'The Seed Blows about in Every Breeze': Noxious Weed Eradication in the Cape Colony, 1860–1909," *Journal of Southern African Studies* 26, no. 4 (2000): 655–74; Brett M. Bennett, "Naturalising Australian Trees in South Africa: Climate, Exotics and Experimentation," *Journal of Southern African Studies* 37, no. 2 (2011): 265–80.

37. Natasha Nongbri, "Plants Out of Place: The 'Noxious Weed' Eradication Campaign in Colonial Southern India," *Indian Economic & Social History Review* 53, no. 3 (2016): 343–69, 346.

38. William Beinart and Karen Middelton, "Plant Transfers in Historical Perspective: A Review Article," *Environment and History* 10, no. 1 (February 2004): 8.

39. Daniel Simberloff, "Ecological Imperialism: The Biological Expansion of Europe, 900–1900, Alfred Crosby," *Quarterly Review of Biology* 62, no. 2 (1987): 223.

40. F. Di Castri, "An Ecological Overview of the Five Regions with a Mediterranean Climate," in *Biogeography of Mediterranean Invasions,* edited by R. H. Groves and F. Di Castri (Cambridge: Cambridge University Press, 1991), 1–16; F. Di Castri, "History of Biology Invasions with Special Emphasis on the Old World," in *Biology Invasions: A Global Perspective,* edited by J. A. Drake et al. (Chichester, UK: Wiley, 1989), 1–29.

41. Di Castri, "History of Biology Invasions," 8.

42. Richard H. Groves, "The Biogeography of Mediterranean Plant Invasions," in *Biogeography of Mediterranean Invasions,* edited by Richard H. Groves and F. Di Castri (Cambridge: Cambridge University Press), 427–38; V. H. Heywood, "Patterns, Extents and Modes of Invasion by Terrestrial Plants," in *Biology Invasions: A Global Perspective,* edited by J. A. Drake et al. (Chichester, UK: Wiley, 1989), 31–60; P. Pyšek, "Is There a Taxonomic Pattern to Plant Invasions?," *Oikos* 82 (1988): 282–94; Jason Fridley, "Of Asian Forests and European Fields: Eastern U.S. Plant Invasions in a Global Floristic Context," *PloS One* 3, 11 (2008): e3630.

43. Daniel Simberloff, *Invasive Species: What Everyone Needs to Know?* (Oxford: Oxford University Press, 2013), 29.

44. Ironically, Fridley in "Of Asian Forests and European Fields" agrees with Crosby but misrepresents Crosby's actual arguments by suggesting that Crosby did not give weight to biogeography, which he certainly did, or to invasions from other parts of Eurasia.

45. Simberloff, *Invasive Species*, 29–30.

46. Simberloff, *Invasive Species*, 36.

47. Daniel Simberloff and Betsy Von Holle, "Positive Interactions of Nonindigenous Species: Invasional Meltdown?," *Biological Invasions* 1, no. 1 (1999): 21–32.

48. Simberloff and Von Holle, "Positive Interactions," 22.

49. Cang Hui and David M. Richardson, *Invasion Dynamics* (Oxford: Oxford University Press, 2017), 8, 12.

50. See Fridley, "Of Asian Forests and European Fields."

51. For grasses, see Veron Visser, John R. U. Wilson, Lyn Fish, Carly Brown, Garry D. Cook, and David M. Richardson, "Much More Give than Take: South Africa as a Major Donor but Infrequent Recipient of Invasive Non-Native Grasses," *Global Ecology and Biogeography* 25, no. 6 (2016): 679–92. For trees, see Marcel Rejmánek, "Invasive Trees and Shrubs: Where Do They Come From and What We Should Expect in the Future?," *Biological Invasions* 16, no. 3 (2014): 483–98.

52. James T. Carlton, "Pattern, Process, and Prediction in Marine Invasion Ecology," *Biological Conservation* 78, no. 1–2 (October–November 1996).

53. W. Rabitsch, Piero Genovesi, Riccardo Scalera, Katarzyna Biała, Melanie Josefsson, and Franz Essl, "Developing and Testing Alien Species Indicators for Europe," *Journal for Nature Conservation* 29 (2016): 89–96. Also see "Invasive Alien Species in Europe: New Framework Shows Scale and Impact Increasing," 451, http://ec.europa.eu/environment/integration/research/newsalert/pdf/invasive_alien_species_europe_new_framework_shows_scale_impact_increasing_451na4_en.pdf (March 2016).

54. Mathieu Rouget, Mark P. Robertson, John R. U. Wilson, Cang Hui, Franz Essl, Jorge L. Renteria, and David M. Richardson, "Invasion Debt—Quantifying Future Biological Invasions," *Diversity and Distributions* 22, no. 4 (2016): 445–56, https://onlinelibrary.wiley.com/doi/abs/10.1111/ddi.12408.

55. Elton is often seen as the intellectual "founder" of invasion biology, even though the modern field developed primarily in the 1980s and 1990s. See Dave Richardson, ed., *Fifty Years of Invasion Ecology: The Legacy of Charles Elton* (Chichester, UK: Wiley-Blackwell, 2011). The work of Matthew Chew has situated Elton's work and ideas within the context of the post–World War II Cold War era. Matthew Chew, "Ending with Elton: Preludes to Invasion Biology" (PhD diss., Arizona State University, 2006). See Matthew Chew, "A Picture Worth Forty-One Words: Charles Elton, Introduced Species, and the 1936 Admiralty Map of British Empire Shipping," *Journal of Transport History* 35, no. 2 (2014): 225–35.

56. Elton's seven initial examples of biological "explosions" only included two species from Europe, with the other five coming from Asia (2), North America (2), and Africa (1). Elton listed the Chinese mitten crab (*Eriocheir sinensis*), the sea lamprey (*Petromyzon marinus*) from North America, cord-grass (*Spartina townsendii*) from England, muskrat (*Ondatra zibethica*) from North America, European starling (*Sturnus vulgaris*), Chestnut blight (*Endothia parasitica*) from China, and African mosquito (*Anopheles gambiae*).

57. Hanno Seebens et al., "No Saturation in the Accumulation of Alien Species Worldwide," *Nature Communications* 8 (2017).

58. Jo Guldi and David Armitage, *The History Manifesto* (Cambridge: Cambridge University Press, 2014).

59. William Beinart, *The Rise of Conservation in South Africa: Settlers, Livestock, and the Environment 1770–1950* (Oxford: Oxford University Press 2003), 376.

60. Beinart, *The Rise of Conservation in South Africa*, 376.

61. Simon Pooley, J. A. Mendelsohn, E. J. Milner-Gulland et al., "Hunting Down the Chimera of Multiple Disciplinary," *Conservation Biology* 28, no. 1 (February 2014): 22–32.

62. Simon Pooley, *Burning Table Mountain: An Environmental History of Fire on the Cape Peninsula* (Claremont, South Africa: UCT Press, 2015); Bella S. Galil, Agnese Marchini, and Anna Occhipinti-Ambrogi, "*Mare Nostrum, Mare Quod Invaditur*—The History of Bioinvasions in the Mediterranean Sea," in *Histories of Bioinvasions in the Mediterranean*, edited by Ana Isabel Queiroz and Simon Pooley, vol. 8, *Environmental History* (2018); see Jane Carruthers and Libby Robin, "Taxonomic Imperialism in the Battles for *Acacia*: Identity and Science in South Africa and Australia," *Transactions of the Royal Society of South Africa* 65, no. 1 (2010): 48–64, among others.

63. J. Fisher, A. D. Manning, W. Steffen, D. B. Rose, K. Daniell, A. Felton, and S. Garnett, "Minding the Sustainability Gap," *Trends in Ecology & Evolution* 22 (2007): 621–24.

64. Gaddis Lewis, *The Landscape of History: How Historians Map the Past* (Oxford: Oxford University Press, 2004), 63.

65. Brett Bennett and Gregory Barton, "Generalizations in Global History: Dealing with Diversity without Losing the Big Picture," *Itinerario* 41, no. 1 (2017): 15–25, 16–18.

66. See Brett Bennett and Joseph Hodge, eds., *Science and Empire: Knowledge and Networks of Science across the British Empire, 1800–1970* (Basingstoke, UK: Palgrave Macmillan, 2011).

67. Beinart and Middelton, "Plant Transfers," offers perhaps the most coherent response but this still does not posit a coherent global theory.

68. See Gaddis, *The Landscape of History*, 64. There is a huge variety of ways historians conceive of causation in history, but the salient point is that scholars criticizing Crosby have generally adhered to the positions described. Scholars in a variety of fields, especially big history, deep history, and environmental history, give agency and causation to physical and biological forces.

69. Bennett and Barton, "Generalizations in Global History," 16–18.

70. For a discussion of the meanings of network, see Bennett and Hodge, *Science and Empire*. For research utilizing a networked approach in this sense, see Jodi Frawley, "Prickly Pear Land: Transnational Networks in Settler Australia," *Australian Historical Studies* 38, no. 130 (2007): 373–88; Frawley, "A Lucky Break: Contingency in the Storied Worlds of Prickly Pear," *Continuum* 28, no. 6 (2014): 760–73.

71. Bruno Latour, "Why Critique Has Run Out of Steam: From Matters of Fact to Matters of Concern," *Critical Inquiry* 30, no. 2 (2004): 225–48, 227.

Contributors

BRETT M. BENNETT is professor in history at the University of Johannesburg and Western Sydney University. He has published widely on various aspects of global environmental history. His books include *Plantations and Protected Areas: A Global History of Forest Management* (2015).

SEMIH CELIK received his PhD from the European University Institute in Florence. He is an expert on the social and environmental history of the Ottoman Empire in the nineteenth century.

NICOLE Y. CHALMER is affiliated with the University of Western Australia. Her research interests focus on the eco-environmental history of food production and ecological interactions between nonhumans and humans, and their cultures and landscapes.

JODI FRAWLEY is an environmental historian based in Brisbane, Australia. She has published widely on different aspects of Australian environmental history. Her research interests include the history of the Royal Botanic Gardens Sydney, the history of invasive species, the history of river environments, and the fishing communities in the Murray-Darling basin.

ULRIKE KIRCHBERGER is a research fellow at the University of Kassel. She has published on different aspects of global and colonial history from the eighteenth to the twentieth century. Her current research deals with the ecological networks and transfers between Australia, South Asia, and Africa from 1850–1920.

CAREY MCCORMACK is an instructor in world history at the University of Tennessee at Chattanooga in the United States. Her research explores indigenous plant knowledge and the development of botany in South and Southeast Asia during the nineteenth century.

IDIR OUAHES is a lecturer in international relations and history at Marbella International University Centre, Spain. He obtained his PhD degree in history from the University of Exeter in 2016. His dissertation was published as *Syria and Lebanon under the French Mandate: Cultural Imperialism and the Workings of Empire* (2018).

FLORIAN WAGNER is assistant professor in contemporary history at the University of Erfurt. He earned his PhD from the European University Institute in Florence, with a dissertation entitled "Colonial Internationalism: How Cooperation among Experts Reshaped Colonialism (1830s–1950s)."

SAMUEL ELEAZAR WENDT is a PhD candidate in history at Europa-Universität Viadrina Frankfurt (Oder), Germany. His research interests include the history of tropical botany and the extraction of colonial cash crops for industrial purposes, transnational history, and commodity chain analysis.

ALEXANDER VAN WICKEREN is affiliated with the University of Cologne. His research interests include the history of science and knowledge from the late eighteenth to the mid-twentieth centuries. He is particularly interested in agricultural knowledge and knowledge of goods and materials.

STEPHANIE ZEHNLE is professor of non-European history at the University of Kiel. She has published widely on the Islamic history of early nineteenth-century West Africa and on human–animal relations in Africa in the nineteenth and twentieth centuries.

Index